Dictionary of the
Environment

NTC POCKET REFERENCES

Dictionary of the
Environment

NTC Publishing Group
Lincolnwood, Illinois USA

Cataloging in Publication Data
is available from the United States Library of Congress

© 1997 by NTC Publishing Group, 4255 West Touhy Avenue,
Lincolnwood (Chicago), Illinois 60646-1975 U.S.A.
Original copyright © 1994 Helicon Publishing Ltd.,
Oxford, England.
Manufactured in the United Kingdom.

Contents

Editorial director
Michael Upshall

Consultant editor
Stephen Webster

Contributors
Simon Fairlie
Sandy Irvine, RealWorld
Nick Middleton

Project editor
Catherine Thompson

Text editor
Paul Davis

Art editor
Terence Caven

Page make-up
Helen Bird

Production
Tony Ballsdon

A

abiotic factor a nonorganic variable within the ecosystem, affecting the life of organisms. Examples include temperature, light, and soil structure. Abiotic factors can be harmful to the environment, as when sulphur dioxide emissions from power stations produce acid rain.

acclimation or *acclimatization* the physiological changes induced in an organism by exposure to new environmental conditions. When humans move to higher altitudes, for example, the number of red blood cells rises to increase the oxygen-carrying capacity of the blood in order to compensate for the lower levels of oxygen in the air.

In evolutionary terms, the ability to acclimate is an important adaptation as it allows the organism to cope with the environmental changes occurring during its lifetime.

accumulation the addition of snow and ice to a glacier. Snow is added through snowfall and avalanches, and is gradually compressed to form ice as the glacier progresses. Although accumulation occurs at all parts of a glacier, it is most significant at higher altitudes near the glacier's start where temperatures are lower. *Ablation* is the loss of snow and ice from a glacier.

acid compound that, in solution in an ionizing solvent (usually water), gives rise to hydrogen ions (H^+ or protons). Acids are defined as substances that are proton donors and accept electrons to form ionic bonds. Acids react with bases to form salts, and they act as solvents. Strong acids are corrosive; dilute acids have a sour or sharp taste, although in some organic acids this may be partially masked by other flavour characteristics.

acid rain acidic deposition caused principally by the pollutant gases sulphur dioxide (SO_2) and the nitrogen oxides. Sulphur dioxide is formed by the burning of fossil fuels, such as coal, that contain high quantities of sulphur; nitrogen oxides are contributed from various industrial activities and from car exhaust fumes. Acid rain is linked with damage to and the death of forests and lake organisms in Scandinavia, Europe, and eastern North America. It also causes damage to buildings and statues.

Even the cleanest rain is slightly acidic (measuring pH 6), owing to the presence of dissolved carbon dioxide. However, as sulphur dioxide dissolves readily in rain drops to form sulphuric acid, the acidity can increase by as much as a thousand times. Acid deposition occurs not only as wet precipitation (mist, snow, or rain), but also comes out of the atmosphere as dry particles or is absorbed directly by lakes, plants, and masonry as gases. Acidic gases can travel over 500 km/300 mi a day so acid rain can be considered an example of transboundary pollution.

The main effect of acid rain is to damage the chemical balance of soil, causing leaching of important minerals including magnesium and aluminium. Plants living in such soils, particularly conifers, suffer from mineral loss and become more prone to infection. The minerals from the soil pass into lakes and rivers, disturbing aquatic life, for instance by damaging the gills of young fish. Lakes and rivers suffer more direct damage as well because they become acidified by rainfall draining directly from their drainage basin.

As a result, thousands of lakes throughout Europe, especially in Scandinavia, have had their food webs seriously disrupted and are now barren of life. Adding alkalis to the lakes in the form of lime neutralizes the acidity, but treatment must be frequent and in itself has some side effects.

US and European power stations that burn fossil fuels release some 8 g/0.3 oz of sulphur dioxide and 3 g/0.1 oz of nitrogen oxides per kilowatt-hour. In 1986, Britain emitted 1,937,000 tonnes of nitrogen oxides to the atmosphere, 40% from power stations and 40% from cars. According to figures from the UK Department of the Environment, emissions of sulphur dioxide from power stations would have to be decreased by 81% in order to arrest damage. The most acidic rain recorded in the UK fell in Pitlochry, Scotland, in 1974, measuring pH 2.4 – the acidity of vinegar.

The only viable solution to acid rain is the reduction of emissions of pollutant gases and international limits have been set by the European Commission. Reductions of UK emissions are being sought by using ⇨flue-gas desulphurization plants in power stations and by fitting more efficient burners; by using gas instead of coal as a power station fuel; and, with road transport rapidly becoming recognized as the single most important source of air pollution, the compulsory fitting of ⇨catalytic converters to all new vehicles.

additive in food, any natural or artificial chemical added to prolong the shelf life of processed foods (salt or nitrates), alter the colour or flavour of food, or improve its food value (vitamins or minerals). Chemical additives are widely used but are subject to regulation, since individuals may be affected by constant exposure even to traces of certain additives and may suffer side effects ranging from headaches and hyperactivity to cancer. Food companies in many countries are now required by law to list additives used in their products. Within the European Community, approved additives are given an official ⇨E number. Organic wholefoods, which contain no added chemicals, are increasing their share of the market (see ⇨organic farming).

Flavours are said to increase the appeal of the food and are used to alter or intensify a food's taste. They may be natural or artificial, and include artificial ⇨sweeteners.
Colourings are used to enhance the visual appeal of certain foods.
Enhancers are used to increase or reduce the taste and smell of a food without imparting a flavour of their own – for example, monosodium glutamate.
Nutrients replace or enhance food value. Minerals and vitamins are added if the diet might otherwise be deficient, to prevent diseases such as beri-beri and pellagra.
Preservatives are antioxidants and antimicrobials that control natural oxidation and the action of microorganisms. They slow down the rate of spoilage by controlling the growth of bacteria and fungi. See ⇨food technology.
Emulsifiers and *surfactants* regulate the consistency of fats in the food and on the surface of the food in contact with the air. They modify the

texture of food and prevent the ingredients of a mixture from separating out.

Thickeners, primarily vegetable gums, regulate the consistency of food. Pectin, for example, is used in this way on fruit products.

Bleaching agents assist in the ageing and whitening of flours.

Antioxidants prevent fatty foods from going rancid by inhibiting their natural oxidation.

advanced gas-cooled reactor (AGR) type of ⇨nuclear reactor widely used in W Europe. The AGR uses a fuel of enriched uranium dioxide in stainless-steel cladding and a moderator of graphite. Carbon dioxide gas is pumped through the reactor core to extract the heat produced by the ⇨fission of the uranium. The heat is transferred to water in a steam generator, and the steam drives a turbogenerator to produce electricity. The AGR is considered superior to its predecessor, the Magnox Reactor, in spite of some early problems. The first AGR reactor in the UK, Dungeness B, was ordered 1965 for service in 1970 but did not begin to operate until 1982 and was still not fully operational 1991.

aerobic a description of those living organisms that require oxygen (usually dissolved in water) for the efficient release of energy contained in food molecules, such as glucose. They include almost all living organisms (plants as well as animals). Those organisms that do not need oxygen for energy release are ⇨anaerobic.

Freshwater systems lose their oxygen when they become polluted, usually because of ⇨algal bloom or the excessive growth of ⇨bacteria. Levels of pollution can often be estimated by observing populations of aerobic organisms, in particular fish, and certain invertebrates such as dragonfly larvae.

aerogenerator wind-powered electricity generator. These range from large models used in arrays on wind farms (see ⇨wind turbine) to battery chargers used on yachts.

aerosol fine suspension of droplets or particles of a liquid or solid floating in a gas, as in fog. Aerosol cans, containing products such as paint, perfume, or deodorant, use pressurized gas as a propellant to force the product through a nozzle in a fine spray.

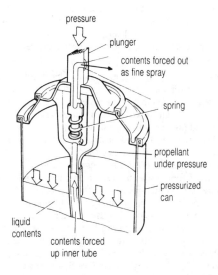

aerosol The aerosol can produces a fine spray of liquid particles, called an aerosol. When the top button is pressed, a valve is opened, allowing the pressurized propellant in the can to force out a spray of the liquid contents. As the liquid sprays from the can, the small amount of propellant dissolved in the liquid vaporizes, producing a fine spray of small droplets.

Most aerosols used chlorofluorocarbons (CFCs) as propellants throughout the 1960s and 1970s because of their apparently useful properties – they were considered to be cheap, convenient, odourless, nonflammable, and nontoxic. However, during the 1980s it was discovered that CFCs cause damage to the ⇨ozone layer in the stratosphere and they are now being phased out in most countries.

The international community has agreed to phase out the use of CFCs, but most so-called 'ozone-friendly' aerosols also use ozone-depleting chemicals, related to CFCs, although not as destructive. Aerosol cans are now generally associated with environmental destruction in the public mind and so are losing some of their former popularity. Some of the products sprayed, such as pesticides, can themselves be directly toxic to humans.

afforestation planting of trees in areas that have not previously held forests. (*Reafforestation* is the planting of trees in deforested areas.) Trees may be planted (1) to provide timber and wood pulp; (2) to provide firewood in countries where this is an energy source; (3) to bind soil together and prevent soil erosion; and (4) to act as windbreaks.

Since 1947 the area of forest in the UK has increased by 70%. In 1991, woodland cover was estimated at 2.3 million hectares, around 10% of total land area. This huge increase has been achieved mainly by planting conifers for commercial use on uplands or on land difficult to farm. In spite of such massive afforestation, around 90% of the UK's timber needs are imported at an annual cost of some £4 billion.

Afforestation is a controversial issue because while many ancient woodlands of mixed trees are being lost, the new plantations consist almost exclusively of conifers. It is claimed that such plantations acidify the soil, conflict with the interests of ⇨biodiversity (they replace more ancient and ecologically valuable species and do not sustain wildlife), and are a blot on previously attractive natural landscapes. A vigorous campaign has been running for many decades against the UK Forestry Commission's afforestation programme in the Lake District on just such grounds.

Afghanistan *the economy and the environment have been devastated by 15 years of civil war following the Soviet invasion 1979. The Soviet 'scorched-earth' policy combined with increasing demands for fuel threatens large areas of the country with deforestation and an estimated 95% of the urban population is without access to sanitation services.*

Agent Orange selective ⇨weedkiller, notorious for its use in the 1960s during the Vietnam War by US forces to eliminate ground cover which could protect enemy forces. It was subsequently discovered to contain highly poisonous ⇨dioxin. Thousands of US troops who had handled it later developed cancer or fathered deformed babies and it had similar impact on Vietnamese victims.

Agent Orange, named after the distinctive orange stripe on its packaging, combines equal parts of 2,4-D (2,4-dichlorophenoxyacetic acid) and 2,4,5-T (2,4,5-trichlorophenoxyacetic acid), both now banned in the USA. Companies that had manufactured the chemicals faced an increasing number of lawsuits in the 1970s. All the suits were settled out of court in a single class action, resulting in the largest ever payment of its kind ($180 million) to claimants.

AGR abbreviation for ⇨*advanced gas-cooled reactor*, a type of nuclear reactor.

agribusiness commercial farming on an industrial scale, often financed by companies whose main interests lie outside agriculture; for example, multinational corporations. Agribusiness farms are mechanized, large in size, highly structured, reliant on chemicals, and are sometimes described as 'food factories'.

agriculture the practice of farming, including the cultivation of the soil (for raising crops) and the raising of domesticated animals. Crops are for human nourishment, animal fodder, or commodities such as cotton and sisal. Animals are raised for wool, milk, leather, dung (as fuel), or meat. The units for managing agricultural production vary from small holdings and individually owned farms to corporate-run farms and collective farms run by entire communities.

Agriculture developed in the Middle East and Egypt at least 10,000 years ago. Farming communities soon became the base for society in China, India, Europe, Mexico, and Peru, then spread throughout the world. Reorganization along more scientific and productive lines took place in Europe in the 18th century in response to dramatic population growth.

Mechanization made considerable progress in the USA and Europe during the 19th century. After World War II, there was an explosive growth in the use of agricultural chemicals: herbicides, insecticides, fungicides, and fertilizers. In the 1960s high-yielding species were developed, especially in the ⇨green revolution of the Third World, and the industrialized countries began intensive farming of cattle, poultry, and pigs. In the 1980s, hybridization by ⇨genetic engineering methods and pest control by the use of chemicals plus pheromones were developed. However, there was also a reaction against some forms of intensive agriculture because of the pollution and habitat destruction caused. One result of this was a growth of alternative methods, including ⇨organic farming.

The total area of land in agricultural use in the UK in 1990 was 18.5 million hectares (around 75% of land in the UK) compared with 19 million hectares in 1980. Of this area, 60% is permanent grassland, the rest mostly arable.

plants For plant products, the land must be prepared by ploughing, cultivating, harrowing, and rolling. Seed must be planted and the growing plant nurtured. This may involve fertilizers, irrigation, pest control by chemicals, and monitoring of acidity or nutrients. When the crop has grown, it must be harvested and, depending on the crop, processed in a variety of ways before it is stored or sold.

Greenhouses allow cultivation of plants that would otherwise find the climate too harsh. Hydroponics allows commercial cultivation of crops using nutrient-enriched solutions instead of soil. Special methods, such as terracing, may be adopted to allow cultivation in hostile terrain and to retain topsoil in mountainous areas with heavy rainfall.

livestock Animals may be semi-domesticated, such as reindeer, or fully domesticated but nomadic (where naturally growing or cultivated food supplies are sparse), or kept in one location. Animal farming involves accommodation (buildings, fencing, or pasture), feeding, breeding, gathering the produce (eggs, milk, or wool), slaughtering, and then further processing such as tanning.

organic agriculture From the 1970s there has been a movement towards using more sophisticated natural methods without chemical sprays and fertilizers. These methods are desirable because nitrates have been seeping into the groundwater; insecticides are found in lethal concentrations at the top of the ⇨food chain; some herbicides are associated with human birth defects, and hormones fed to animals to promote fast growth have damaging effects on humans. See also ⇨organic farming.

overproduction The greater efficiency in agriculture achieved since the

19th century, coupled with post–World War II government subsidies for domestic production in the USA and the European Community (EC), have led to the development of high stocks, nicknamed 'lakes' (wine, milk) and 'mountains' (butter, beef, grain). There is no simple solution to this problem, as any large-scale dumping onto the market displaces regular merchandise and so may drive out many small uneconomic producers who, by switching to other enterprises, risk upsetting the balance elsewhere. Increasing concern about the starving and the cost of storage has led the USA and the EC to develop measures for limiting production, such as letting arable land lie fallow to reduce grain crops (set-aside) and quotas for production (for example, milk). The USA had some success at selling surplus wheat to the USSR when the Soviet crop was poor, but the overall cost of bulk transport and the potential destabilization of other economies has acted against high producers exporting their excess on a regular basis to needy countries. Intensive farming methods also contribute to soil ⇨erosion and water pollution.

agricultural chemicals After World War II, there was an explosive growth in the use of agricultural chemicals, or agrochemicals, such as herbicides, insecticides, fungicides, and fertilizers. In the 1950s and 1960s, high-yield varieties of crops such as maize and rice were developed, which increased food production in the Third World (the ⇨green revolution) but increased reliance on agrochemicals. At the same time the industrialized countries began intensive farming of cattle, poultry, and pigs with the introduction of genetic engineering and selective breeding. From the 1970s concern about the damage caused to the environment and to health by some farming methods has led to a movement toward organic methods of production (see above).

Nitrates in fertilizers can leach away from the soil to pollute water supplies, and pesticides can pass through food chains, accumulating in the diets of animals, including humans. The intensive rearing (⇨factory farming) of animals such as pigs and chickens has lowered the cost of meat, but has also aroused controversy about its cruelty and about possible health hazards such as salmonella food poisoning. Land clearance and ⇨deforestation have destroyed the natural habitats of many animal and plant species and has also led to soil erosion in which the top, fertile layer of soil – no longer anchored by tree and shrub roots – is blown or washed away, leaving behind a barren desert or '⇨dust bowl'.

agrochemical artificially produced chemical used in modern, intensive agricultural systems. Agrochemicals include nitrate and phosphate fertilizers, pesticides, some animal-feed additives, and pharmaceuticals. Many are responsible for pollution and almost all are avoided by organic farmers.

aid money or resources given or lent on favourable terms to poorer countries. A distinction may be made between *short-term aid* (usually food and medicine), which is given to relieve conditions in emergencies such as famine, and *long-term aid*, which is intended to promote economic activity and improve the quality of life – for example, by funding irrigation, education, and communications programmes.

In the late 1980s official aid from governments of richer nations amounted to $45–$60 billion annually, whereas voluntary organizations, such as Oxfam, received about $2.4 billion a year for the developing countries. The World Bank is the largest dispenser of aid. All industrialized United Nations member countries devote a proportion of their gross national product (GNP) to aid, ranging from 0.2% of GNP (Ireland) to 1.1% (Norway).

In the UK, the Overseas Development Administration (ODA) is the department of the Foreign Office that handles bilateral aid (aid from the UK by agreement with another country). The combined overseas development aid of all European Community (EC) member countries is less than the sum ($20 billion) the EC spends every year on storing surplus food produced by European farmers. The UK development-aid budget in 1988 was 0.32% of GNP, with India and Kenya among the principal beneficiaries.

air mass large body of air with particular characteristics of temperature and humidity. An air mass forms when air rests over an area long enough to pick up the conditions of that area. For example, an air mass formed over the Sahara will be hot and dry. When an air mass moves to another area it affects the ⇨climate of that area, but its own characteristics become modified in the process. For example, a Saharan air mass becomes cooler as it moves northwards. Air masses can carry airborne pollution such as acid rain at speeds of up to 500 km/300 mi a day.

air mass

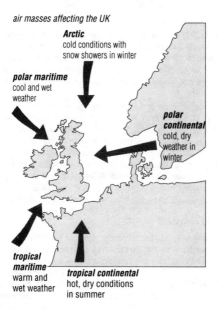

air masses affecting the UK

Arctic
cold conditions with
snow showers in winter

polar maritime
cool and wet
weather

**polar
continental**
cold, dry
weather in
winter

**tropical
maritime**
warm and
wet weather

tropical continental
hot, dry conditions
in summer

air pollution contamination of the atmosphere caused by the discharge, accidental or deliberate, of a wide range of toxic airborne substances. Often the amount of the released substance is relatively high in a certain locality, so the harmful effects become more noticeable. The cost of preventing any discharge of pollutants into the air is prohibitive, so attempts are more usually made to gradually reduce the amount of discharge and to disperse this as quickly as possible by using very tall chimneys, or by intermittent release.

The greatest single cause of air pollution in the UK is the ⇨car, which is responsible for 85% of the carbon monoxide and 45% of the oxides of nitrogen present in the atmosphere.

IS AIR TRAVEL ENVIRONMENTALLY UNACCEPTABLE?

Air travel is one of the world's fastest-growing energy users, and it is also potentially one of our most environmentally damaging activities. Although the environmental impact of aircraft is generally equated with the problems of noise and perhaps with fuel consumption, it now seems likely that the most significant impact is the effect that airliners and military aircraft have on the ozone layer, and on global warming.

The ozone layer is a thin but vital gaseous layer in the upper atmosphere (stratosphere), which helps prevent harmful solar radiation from reaching the Earth. It is currently thinning in places, owing to the effects of several chemically stable pollutants that reach the stratosphere and destroy ozone molecules. Global warming is a complex series of known and suspected climatic effects that are occurring because a greater amount of solar heat is being trapped inside the Earth's atmosphere as a result of increased levels of pollutants, including carbon dioxide, methane, and nitrogen oxides.

Increased demand

Flying is now seen as virtually indispensable for many people in the wealthy countries of the North, and for the rich minority in the South. In 1989, scheduled services carried more than one billion passengers worldwide, with about a quarter of the flights being international. Over the past five years, passenger numbers have increased by 5% a year on average, with a proportionately greater increase in international flights. The main reasons for air travel differ between countries, but holidays and pleasure trips make up an increasing proportion of journeys in the richest countries. In the UK, for example, only about 20% of 1989 flights were for business, with the remainder being divided between package holidays (38%), other holidays, visiting relatives, and so on.

Flying is the standard way of getting around countries where internal distances are especially large, such as the USA and Australia. The International Civil Aviation Authority expects this rate of increase to continue for some time, which means that the demand for air transport would approximately double by 2005.

In general, pollution emissions from aircraft form a relatively small proportion of total global pollution levels. However, the pollutants are released at very high altitudes, and thus injected directly into the stratosphere, where they may do more harm. Pollutants persist longer at high altitudes than they do near the ground, and some important atmospheric chemical reactions, such as ozone formation, occur in the stratosphere. The impact that aircraft could have on ozone is complex, and varies with a number of factors, including altitude. At medium altitudes (the troposphere), the nitrogen oxide (NOx) emissions from aircraft could help *produce* ozone, which, at that height above the planet's surface, has a significant global-warming potential. However, in the stratosphere NOx could have the opposite effect – *destroying* ozone. In addition, high-level clouds could be formed in part by water emissions from aircraft. Both these factors would help increase the net influx of radiation and thus increase global warming. Ozone loss has a number of other detrimental effects, including causing skin cancer and cataracts.

What can be done?

Despite the potential seriousness of the problem, there has been comparatively little detailed research on the impact of aircraft on global warming. The various different effects make firm predictions even more difficult than is usually the case in atmospheric chemistry. A report published by the World Wide Fund for Nature (WWF) in 1991 suggested that the contribution made by aircraft to global warming could be anything between 2% and 43% of the total. This means that the global-warming potential of aircraft may be anything from small but significant to so large that up to a quarter of our efforts at reducing global warming should be directed at aircraft.

That said, it is difficult to see what can be done to reduce the impact in the short term except to persuade people to cut down on air travel. The development of high-speed trains and other improved forms of land transport, as well as the increasing sophistication of communication networks, could help reduce business travel in the future. But with the bulk of travel being for pleasure in many countries, pleas for a reduction in flying are likely to meet with entrenched opposition. Future research will show how important air transport really is in terms of global warming. If the pessimistic figures are correct, a major rethink of air-transport costs and availability seems inevitable.

Alaska *a Congressional act 1980 gave environmental protection to 42 million hectares. Valuable mineral resources have been exploited from 1968, especially in the Prudhoe Bay area to the SE of Point Barrow. An oil pipeline (1977) runs from Prudhoe Bay to the port of Valdez. An oil-spill from a tanker in Prince William Sound caused great environmental damage 1989. An underground natural-gas pipeline to Chicago and San Francisco is currently under construction.*

algae (singular *alga*) diverse group of plants (including those commonly called seaweeds) that shows great variety of form, ranging from single-celled forms to multicellular seaweeds of considerable size and complexity. Marine algae help combat global warming by removing carbon dioxide from the atmosphere during photosynthesis.

algal bloom The sudden appearance, in a lake or river, of a vast population of algae, typically giving the water a red or green coloration.

Such blooms may be natural but are often the result of nitrate pollution, in which artificial fertilizers, applied to surrounding fields, leach out into the waterways. This type of bloom can lead to the death of almost every other organism in the water; because light cannot penetrate the algal growth, the plants beneath can no longer photosynthesize and therefore do not release oxygen into the water. Only those organisms that are adapted to very low levels of oxygen survive.

alkali a compound classed as a base that is soluble in water. Alkalis neutralize acids and are soapy to the touch. The hydroxides of metals are alkalis: those of sodium (sodium hydroxide, $NaOH$) and of potassium (potassium hydroxide, KOH) being chemically powerful; both were derived from the ashes of plants.

alpha particle positively charged, high-energy particle emitted from the nucleus of a radioactive atom. It is one of the products of the spontaneous disintegration of radioactive elements (see ⇨radioactivity) such as radium and thorium, and is identical with the nucleus of a helium atom – that is, it consists of two protons and two neutrons. The process of emission, *alpha decay*, transforms one element into another, decreasing the atomic (or proton) number by two and the atomic mass (or nucleon number) by four.

Because of their large mass, alpha particles have a short range of only a few centimetres in air, and can be stopped by a sheet of paper. They have a strongly ionizing effect (see ⇨ionizing radiation) on the molecules that they strike, and are therefore capable of damaging living cells. Alpha particles travelling in a vacuum are deflected slightly by magnetic and electric fields.

alternative energy see ⇨energy, alternative

aluminium lightweight, silver-white, ductile and malleable, metallic element, symbol Al, atomic number 13, relative atomic mass 26.9815. It is the third most abundant element (and the most abundant metal) in the Earth's crust, of which it makes up about 8.1% by mass. It is an excellent conductor of electricity and oxidizes easily, the layer of oxide on its surface making it highly resistant to tarnish.

Because of its rapid oxidation a great deal of energy is needed in order to separate aluminium from its ores, and the pure metal was not readily obtainable until the middle of the 19th century. Commercially, it is prepared by the electrolysis of ⇨bauxite.

Aluminium sulphate is the most widely used chemical in water treatment worldwide, but accidental excess (as at Camelford, N Cornwall, UK, July 1989) makes drinking water highly toxic, and discharge into rivers kills all fish.

Amazon South American river, the world's second longest, 6,570 km/4,080 mi, and the largest in volume of water. It has 48,280 km/30,000 mi of navigable waterways, draining 7,000,000 sq km/2,750,000 sq mi, nearly half the South American landmass. It reaches the Atlantic on the equator, its estuary 80 km/50 mi wide, discharging a volume of water so immense that 64 km/40 mi out to sea fresh water remains at the surface. The Amazon basin covers 7.5 million sq km/3 million sq mi, of which 5 million sq km/2 million sq mi is tropical forest containing 30% of all known plant and animal species (80,000 known species of trees, 3,000 known species of land vertebrates, 2,000 freshwater fish). It is the wettest region on Earth; average rainfall 2.54 m/8.3 ft a year.

The opening up of the Amazon river basin to settlers from the over-populated east coast has resulted in a massive burning of tropical forest to create both arable and pasture land. The problems of soil erosion, the disappearance of potentially useful plant and animal species, and the possible impact of large-scale forest clearance on global warming of the atmosphere have become environmental issues of international concern.

Brazil, with one third of the world's remaining tropical rainforest, has 55,000 species of flowering plant, half of which are only found in Brazilian Amazonia. In June 1990 the Brazilian Satellite Research Institute announced that 8% of the rainforest in the area had been destroyed by deforestation, amounting to 404,000 sq km/156,000 sq mi (nearly the size of Sweden).

Amoco Cadiz US-owned oil tanker which ran aground off the French Coast 1978, releasing 225,000 tonnes of crude oil. The environmental impact was immense – 30,000 seabirds died, 130 beaches were immersed in a 30 cm/12 in layer of oil, and 230,000 tonnes of crustacea and fish perished. The fact that the accident took place in Western Europe led to massive media coverage of the environmental damage and so raised awareness of oil pollution.

anaerobic (of living organisms) not requiring oxygen for the release of energy from food molecules such as glucose. Anaerobic organisms include many bacteria, yeasts, and internal parasites.

Obligate anaerobes such as certain primitive bacteria cannot function in the presence of oxygen; but *facultative anaerobes*, like the fermenting yeasts and most bacteria, can function with or without oxygen. Anaerobic organisms release 19 times less of the available energy from their food than do ⇨aerobic organisms.

Although anaerobic respiration is a primitive and inefficient form of energy release, deriving from the period when oxygen was missing from the atmosphere, it can also be seen as an adaptation. To survive in some habitats, such as the muddy bottom of a polluted river, an organism must be to a large extent independent of oxygen; such habitats are said to be *anoxic*.

animal rights the extension of the concept of human rights to animals on the grounds that animals may not be able to reason but can suffer and are easily exploited by humans. The ***animal rights movement*** is a general description for a wide range of organizations, both national and local, that take a more radical approach than the traditional *welfare* societies, such as the RSPCA. More radical still is the concept of *animal liberation*, that animals should not be used or exploited in any way at all.

Although there has long been concern for animals, the animal rights movement became fully established in most Western countries in the late 1970s, particularly following the publication of *Animal Liberation* 1975 by Australian philospher Peter Singer (1946–), which gave the movement a solid theoretical basis. Animal rights activists are best known for campaigns against the ⇨fur and meat trades, ⇨vivisection and bloodsports, but also oppose other forms of animal exploitation such as ⇨zoos and the pet trade. Methods range from leafletting and demonstrations to illegal direct action, such as attacking butchers' shops and removing animals from vivisection laboratories.

Antarctica occupying 10% of the world's surface, Antarctica contains 90% of the world's ice and 70% of its fresh water. The Antarctic ecosystem is characterized by large numbers of relatively few species of higher plants and animals, and a short food chain from tiny marine plants to whales, seals, penguins, and other sea birds. ⇨Whaling, which began around Antarctica in the early 20th century, ceased during the 1960s as a result of overfishing, although Norway and Iceland defied the ban 1992 to recommence whaling. Petroleum, coal, and minerals such as palladium and platinum exist, but their exploitation is prevented by a 50-year ban on commercial mining agreed by 39 nations 1991.

anthracite hard, dense, shiny variety of coal, containing over 90% carbon and a low percentage of ash and impurities, which causes it to burn without flame, smoke, or smell. Because of its purity, anthracite gives off relatively little sulphur dioxide when burnt.

antibiotic drug that kills or inhibits the growth of bacteria and fungi. It is derived from living organisms such as fungi or bacteria, which distinguishes it from synthetic antimicrobials.

The earliest antibiotics, the penicillins, came into widespread use from 1941 and were quickly joined by chloramphenicol, the cephalosporins, erythromycins, tetracyclines, and aminoglycosides. A range of broad-spectrum antibiotics, the 4-quinolones, was developed 1989, of which ciprofloxacin was the first. Each class and individual antibiotic acts in a different way and may be effective against either a broad spectrum or a specific type of disease-causing agent. Use of antibiotics has become more selective as side effects, such as toxicity, allergy, and resistance, have become better understood. Bacteria have the ability to develop immunity following repeated or subclinical (insufficient) doses, so more advanced and synthetic antibiotics are continually required to overcome them.

appropriate technology simple or small-scale machinery and tools that, because they are cheap and easy to produce and maintain, may be of most use in the developing world; for example, hand ploughs and simple looms. This equipment may be used to supplement local crafts and traditional skills to encourage small-scale industrialization.

An important aspect of appropriate technology is its reliance on easily available energy, for example, from small petrol engines, human power, wind pumps, and solar panels. This avoids the need for large capital investment, a scarce resource in the developing world, but utilizes the generally large supplies of human labour.

aquifer any rock formation containing water. The rock of an aquifer must be porous and permeable (full of interconnected holes) so that it can absorb water. Aquifers are an important source of fresh water, for example, for drinking and irrigation, in many arid areas of the world and are exploited by the use of ⇨artesian wells.

An aquifer may be underlain, overlain, or sandwiched between impermeable layers, called *aquicludes*, which impede water movement. Sandstones and porous limestones make the best aquifers.

Argentina *the population is heavily concentrated in major urban centres which has created problems with air and water pollution. Heavy flooding is also a problem near rivers; an estimated 20,000 sq km/7,700 sq mi of land has been swamped with salt water.*

---

arid region a region that is very dry and has little vegetation. Aridity depends on temperature, rainfall, and evaporation, and so is difficult to quantify, but an arid area is usually defined as one that receives less than

250 mm/10 in of rainfall each year. (By comparison, New York City receives 1,120 mm/4.4 in per year and London has an average annual rainfall of 600 mm/2.4 in). There are arid regions in North Africa, Pakistan, Australia, the USA, and elsewhere. Very arid regions are known as deserts.

The scarcity of fresh water is a problem for the inhabitants of arid zones, and constant research goes into discovering cheap methods of distilling sea water and artificially recharging natural groundwater reservoirs. Another problem is the eradication of salt in irrigation supplies from underground sources or where a surface deposit forms in poorly drained areas.

arsenic brittle, greyish-white, semi-metallic element (a metalloid), symbol As, atomic number 33, relative atomic mass 74.92. It occurs in many ores and occasionally in its elemental state, and is widely distributed, being present in minute quantities in the soil, the sea, and the human body. In larger quantities, it is poisonous. The chief source of arsenic compounds is as a by-product from metallurgical processes. It is used in making semiconductors, alloys, and solders.

As it is a cumulative poison, its presence in food and drugs is

artesian well

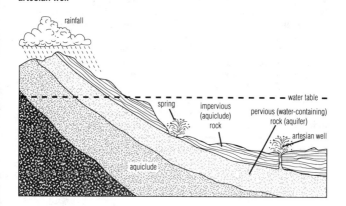

extremely dangerous. The symptoms of arsenic poisoning are vomiting, diarrhoea, tingling and possibly numbness in the limbs, and collapse.

artesian well well that is supplied with water rising under its own pressure from an underground water-saturated rock layer (⇨aquifer). Such a well may be drilled into an aquifer that is confined by impermeable rocks both above and below. If the water table (the top of the region of water saturation) in that aquifer is above the level of the well head, hydrostatic pressure will force the water to the surface.

Because their water is fresh and easily available, artesian wells are often overexploited and eventually become unreliable. There is also some concern that pollutants such as pesticides or nitrates can seep into the aquifers.

artificial radioactivity natural and spontaneous radioactivity arising from radioactive isotopes or elements that are formed when elements are bombarded with subatomic particles – protons, neutrons, or electrons – or small nuclei.

artificial selection selective breeding of individuals that exhibit the particular characteristics that a plant or animal breeder wishes to develop. In plants, desirable features might include resistance to disease, high yield (in crop plants), or attractive appearance. In animal breeding, selection has led to the development of particular breeds of cattle for improved meat production (such as the Aberdeen Angus) or milk production (such as Jerseys).

High-yielding, artificially selected organisms such as beef, cattle, pigs, rice, and wheat often need carefully controlled environments. The increased use of antibiotics and other medicines in animal husbandry, or of pesticides and fertilizers in arable farming, is a controversial consequence of some artificial breeding programmes.

Artificial selection in itself is by no means a product of modern technology. It was practised by the Sumerians at least 5,500 years ago and carried on through the succeeding ages, so that all common vegetables, fruit, and livestock have been long modified by selective breeding.

asbestos any of several related minerals of fibrous structure that offer great heat resistance because of their nonflammability and poor conductivity. Commercial asbestos is generally either made from serpentine ('white' asbestos) or from sodium iron silicate ('blue' silicate). The fibres are woven together or bound by an inert material. Over time the fibres can work loose and, because they are small enough to float freely in the air or be inhaled, asbestos usage is now strictly controlled because exposure to its fibres can cause cancer (mesothelioma) by lodging in the lungs and damaging the cells. *Asbestosis* is a chronic lung inflammation caused by asbestos fibres. Blue asbestos is more dangerous than the more common white asbestos.

atmosphere the protective envelope of gases that surrounds the Earth, prevented from escaping by the pull of the Earth's gravity. Atmospheric pressure decreases with height in the atmosphere. In its lowest layer, the atmosphere consists of nitrogen (78%) and oxygen (21%), both in molecular form (two atoms bounded together). The other 1% is largely argon, with very small quantities of other gases, including water vapour and carbon dioxide. The atmosphere plays a major part in the various cycles of nature (the ⇨water cycle, ⇨carbon cycle, and ⇨nitrogen cycle). It is the principal industrial source of nitrogen, oxygen, and argon, which are obtained by fractional distillation of liquid air.

The lowest level of the atmosphere, the troposphere, is heated by the Earth, which is warmed by infrared and visible radiation from the Sun. Warm air cools as it rises in the *troposphere*, causing rain and most other weather phenomena. However, infrared and visible radiations form only a part of the Sun's output of electromagnetic radiation. Almost all the shorter-wavelength ultraviolet radiation is filtered out by the upper layers of the atmosphere. The filtering process is an active one: at heights above about 50 km/31 mi ultraviolet photons collide with atoms, knocking out electrons to create a plasma of electrons and positively charged ions. The resulting *ionosphere* acts as a reflector of radio waves, enabling radio transmissions to 'hop' between widely separated points on the Earth's surface.

Waves of different wavelengths are reflected best at different heights. The collisions between ultraviolet photons and atoms lead to a heating of the upper atmosphere, although the temperature drops from top to bottom within the zone called the *thermosphere* as high-energy photons are progressively absorbed in collisions. Between the

composition of the atmosphere

gas	symbol	volume (%)	role
nitrogen	N_2	78.08	cycled through human activities and through the action of microorganisms on animal and plant waste
oxygen	O_2	20.94	cycled mainly through the respiration of animals and plants and through the action of photosynthesis
carbon dioxide	CO_2	0.03	cycled through respiration andphotosynthesis in exchange reactions with oxygen. It is also a product of burning fossil fuels
argon	Ar	0.093	chemically inert and with only a few industrial uses
neon	Ne	0.0018	as argon
helium	He	0.0005	as argon
krypton	Kr	trace	as argon
xenon	Xe	trace	as argon
ozone	O_3	0.00006	a product of oxygen molecules split into single atoms by the Sun's radiation and unaltered oxygen molecules
hydrogen	H_2	0.00005	unimportant

thermosphere and the ***tropopause*** (at which the warming effect of the Earth starts to be felt) there is a 'warm bulge' in the graph of temperature against height, at a level called the ***stratopause***. This is due to longer-wavelength ultraviolet photons that have survived their journey through the upper layers; now they encounter molecules and split them apart into atoms. These atoms eventually bond together again, but often in different combinations. In particular, many ⇨ozone molecules (oxygen atom triplets) are formed. Ozone is a better absorber of ultraviolet than ordinary (two-atom) oxygen, and it is the ***ozone layer*** that prevents lethal amounts of ultraviolet from reaching the Earth's surface.

Far above the atmosphere, as so far described, lie the ***Van Allen radiation belts***. These are regions in which high-energy charged particles travelling outwards from the Sun (as the so-called solar wind) have been captured by the Earth's magnetic field. The outer belt (at about

1,600 km/1,000 mi) contains mainly protons, the inner belt (at about 2,000 km/1,250 mi) contains mainly electrons. Sometimes electrons spiral down towards the Earth, noticeably at polar latitudes, where the magnetic field is strongest. When such particles collide with atoms and ions in the thermosphere, light is emitted. This is the origin of the glows visible in the sky as the *aurora borealis* (northern lights) and the *aurora australis* (southern lights). A fainter, more widespread, *airglow* is caused by a similar mechanism.

During periods of intense solar activity, the atmosphere swells outwards; there is a 10–20% variation in atmosphere density. One result is to increase drag on satellites. This effect makes it impossible to predict exactly the time of re-entry of satellites.

atmospheric pollution contamination of the atmosphere with the harmful by-products of human activity; see ⇨air pollution.

atom smallest unit of matter that can take part in a chemical reaction, and which cannot be broken down chemically into anything simpler. An atom is made up of protons and neutrons in a central nucleus surrounded by electrons. The atoms of the various elements differ in atomic number, relative atomic mass, and chemical behaviour. There are 109 different types of atom, corresponding with the 109 known elements as listed in the periodic table of the elements.

atomic energy another name for ⇨nuclear energy.

atomic radiation energy given out by disintegrating atoms during ⇨radioactive decay, whether natural or synthesized. The energy may be in the form of fast-moving particles, known as ⇨alpha particles and ⇨beta particles, or in the form of high-energy electromagnetic waves known as ⇨gamma radiation. Exposure to atomic radiation is linked to chromosomal damage, cancer, and, in laboratory animals at least, hereditary disease. In the UK, it is estimated that about 3% of all cancer deaths may result from the effects of radiation (see also ⇨radiation sickness).

atomic testing the detonation of nuclear devices so as to verify their reliability, power and destructive abilities. Although carried out secretly in remote regions of the world, such tests are easily detected by the seismic shock waves produced. The first tests were carried out in the atmosphere during the 1950s by the USA, the former USSR, and the UK. These tests produced large quantities of radioactive fallout, still active today. A test ban in 1963 prohibited all tests except those carried out underground. Some 2,000 tests have been carried out since World War II, an average of one every nine days.

Australia *an estimated 75% of Australia's northern tropical rainforest has been cleared for agriculture or urban development since Europeans first settled there in the early 19th century. Soil erosion is also a problem due to inapproptiate land use and there are many species unique to the continent that are in danger of extinction due to loss of habitat.*

autotroph any living organism that synthesizes organic substances from inorganic molecules by using light or chemical energy. Autotrophs are the *primary producers* in all ⇨food chains since the materials they synthesize and store are the energy sources of all other organisms. All green plants and many planktonic organisms are autotrophs, using sunlight to convert carbon dioxide and water into sugars by ⇨photosynthesis.

The total ⇨biomass of autotrophs is far greater than that of animals, reflecting the dependence of animals on plants, and the ultimate dependence of all life on energy from the Sun – green plants convert light energy into a form of chemical energy (food) that animals can exploit. It is estimated that 10% of the energy in autotrophs can pass into the next stage of the food chain, the rest being lost as heat or indigestible matter. See also ⇨heterotroph.

aye-aye nocturnal tree-climbing prosimian *Daubentonia madagascariensis* of Madagascar, related to the lemurs. It is just over 1 m/3 ft long, including a tail 50 cm/20 in long. The aye-aye has become increasingly rare due to loss of its forest habitat, and is now classified as an endangered species.

B

background radiation radiation that is always present in the environment. By far the greater proportion (87%) of it is emitted from natural sources. Alpha and beta particles, and gamma radiation are radiated by the traces of radioactive minerals that occur naturally in the environment and even in the human body, and by radioactive gases such as radon and thoron, which are found in soil and may seep upwards into buildings. Radiation from space (cosmic radiation) also contributes to the background level. The remaining 13% is almost entirely medical in origin. Less than 0.1% of background radiation can be traced to the normal operation of nuclear power stations.

backwash the retreat of a wave that has broken on a ⇨beach. When a wave breaks, water rushes up the beach as swash and is then drawn back towards the sea as backwash.

bacteria (singular *bacterium*) microscopic unicellular organisms. They usually reproduce by binary fission (dividing into two equal parts), and since this may occur approximately every 20 minutes, a single bacterium is potentially capable of producing 16 million copies of itself in a day. It is thought that 1–10% of the world's bacteria have been identified.

   Although popularly considered harmful, certain types of bacteria are vital for their role in the decomposition of dead organic matter and as such are an essential component of the ⇨nitrogen cycle, while others are used in many food and industrial processes.

badlands barren landscape cut by erosion into a maze of ravines, pinnacles, gullies, and sharp-edged ridges. Examples include areas in South Dakota and Nebraska, USA.

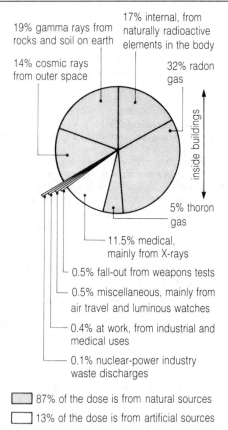

19% gamma rays from rocks and soil on earth

17% internal, from naturally radioactive elements in the body

14% cosmic rays from outer space

32% radon gas

inside buildings

5% thoron gas

11.5% medical, mainly from X-rays

0.5% fall-out from weapons tests

0.5% miscellaneous, mainly from air travel and luminous watches

0.4% at work, from industrial and medical uses

0.1% nuclear-power industry waste discharges

87% of the dose is from natural sources

13% of the dose is from artificial sources

background radiation *The radiation produced by the nuclear power industry is minute compared with the radiation produced naturally by rocks, cosmic rays and radioactive elements within the body.*

Badlands, which can be created by overgrazing, are so called because of their total lack of value for agriculture and their inaccessibility.

balance of nature the idea that there is an inherent equilibrium in most ⇨ecosystems, with plants and animals interacting so as to produce a stable, continuing system of life on Earth. Organisms in the ecosystem are adapted to each other – for example, waste products produced by one species are used by another and resources used by some are replenished by others; the oxygen needed by animals is produced by plants while the waste product of animal respiration, carbon dioxide, is used by plants as a raw material in photosynthesis. The nitrogen cycle, the water cycle, and the control of animal populations by natural predators are other examples. The activities of human beings can, and frequently do, disrupt the balance of nature. This stability is not seen in all natural ecosystems: in ⇨succession, for example, there is progressive change, while some ecosystems are characterized by population cycles with an explosion of numbers being followed by catastrophic decline.

The idea of a balance of nature is also expressed in the ⇨Gaia hypothesis, which likens the Earth to a living organism, constantly adjusting itself to circumstances so as to increase its chances of survival.

Bangladesh *deforestation on the slopes of the Himalayas increases the threat of flooding in the coastal lowlands of Bangladesh, which are also subject to devastating monsoon storms. The building of India's Farakka Barrage has reduced the flow of the Ganges in Bangladesh and permitted salt water to intrude further inland. Increased salinity has destroyed fisheries, contaminated drinking water, and damaged forests.*

---

bar unit of pressure equal to 10^5 pascals or 10^6 dynes/cm^2, approximately 750 mmHg or 0.987 atm. Its diminutive, the ***millibar*** (one-thousandth of a bar), is commonly used by meteorologists.

bar deposit of sand or silt formed in a river channel, or a long sandy ridge running parallel to a coastline. Coastal bars can extend across estuaries to form ***bay bars***.

Barbary ape tailless, yellowish-brown macaque monkey *Macaca* *sylvanus*, found in the mountains and wilds of Algeria and Morocco, especially in the forests of the Atlas mountains. The macaque is threatened by illegal logging, which is devastating some of the ancient forests in the area. Although it is breeding well in captivity, forest loss may confound attempts to reintroduce this species into the wild.

barrel unit of liquid capacity, the value of which depends on the liquid being measured. It is used for petroleum, a barrel of which contains 159 litres/35 imperial gallons.

bauxite principal ore of ⇨aluminium, consisting of a mixture of hydrated aluminium oxides and hydroxides, generally contaminated with compounds of iron, which give it a red colour. Bauxite is formed by the ⇨chemical weathering of rocks in tropical climates. Chief producers of bauxite are Australia, Guinea, Jamaica, Russia, Kazakhstan, Surinam, and Brazil.

To produce aluminium the ore is processed into a white powder (alumina), from which the aluminium is extracted by passing a large electric current through it. The process is only economical if cheap electricity is readily available, usually from a hydroelectric plant.

beach strip of land bordering the sea, normally consisting of boulders and pebbles on exposed coasts or sand on sheltered coasts. It is usually defined by the high- and low-water marks.

The material of the beach consists of a rocky debris eroded from exposed rocks and headlands. The material is transported to the beach, and along the beach, by waves that hit the coastline at an angle, resulting in a net movement of the material in one particular direction. This movement is known as *longshore drift*. Attempts are often made to halt longshore drift by erecting barriers, or jetties, at right angles to the movement. Pebbles are worn into round shapes by being battered against one another by wave action and the result is called *shingle*. The finer material, the *sand*, may be subsequently moved about by the wind and form sand dunes. Apart from the natural process of longshore drift, a beach may be threatened by the commercial use of sand and aggregate, by the mineral industry – since particles of metal ore are often concentrated into workable deposits by the wave action – and by pollution (for example, by oil spilt or dumped at sea).

Concern for the conditions of bathing beaches led in the 1980s to a directive from the European Economic Community on water quality. In the UK, beaches free of industrial pollution, litter, and sewage, and with water of the highest quality, have the right (since 1988) to fly a blue flag.

bear large mammal with a heavily built body, short powerful limbs, and a very short tail. Bears breed once a year, producing one to four cubs. In northern regions they hibernate, and the young are born in the winter den. They are found mainly in North America and N Asia.

The bear family, Ursidae, is related to carnivores such as dogs and weasels, and all are capable of killing prey. (The panda is probably related to both bears and raccoons.) Of the seven species of bear, five are currently reckoned to be endangered and all, apart from the polar bear and the American black bear, are in decline. In 1992, the American black bear was upgraded to Appendix 2 of CITES (Convention on International Trade in Endangered Species) to stem the trade in the species' gall bladder.

beaver aquatic rodent *Castor fiber* with webbed hind feet, a broad flat scaly tail, and thick waterproof fur. It has very large incisor teeth and fells trees to feed on the bark and to use the logs to construct the 'lodge', in which the young are reared, food is stored, and much of the winter is spent.

Beavers can construct dams on streams, and thus modify the environment considerably. They once ranged across Europe, N Asia, and North America, but in Europe now only survive where they are protected, and are reduced elsewhere, partly through trapping for their ⇨fur.

becquerel SI unit (symbol Bq) of ⇨radioactivity, equal to one radioactive disintegration (change in the nucleus of an atom when a particle or ray is given off) per second.

benzene C_6H_6 clear liquid hydrocarbon of characteristic odour, occurring in coal tar. It is used as a solvent and in the synthesis of many chemicals.

The benzene molecule consists of a ring of six carbon atoms, all of which are in a single plane, and it is one of the simplest cyclic compounds. Benzene is the simplest of a class of compounds collectively known as *aromatic compounds*. Some are considered carcinogenic (cancer-inducing) and there is concern about benzene derivatives found in the exhaust fumes of diesel vehicles.

benzpyrene one of a number of organic compounds associated with a particular polycyclic ring structure. Benzpyrenes are present in coal tar at low levels and are considered carcinogenic (cancer-inducing). Traces of benzpyrenes are present in wood smoke, and this has given rise to some concern about the safety of naturally smoked foods.

beta particle electron ejected with great velocity from a radioactive atom that is undergoing spontaneous disintegration. Beta particles do not exist in the nucleus but are created by beta decay, when a neutron inside the nucleus converts to a proton, emitting a beta particle (electron) in the process.

Beta particles are more penetrating than ⇨alpha particles, but less so than ⇨gamma radiation; they can travel several metres in air, but are stopped by 2–3 mm of aluminium. They are less strongly ionizing than alpha particles and, like cathode rays, are easily deflected by magnetic and electric fields.

bicycle pedal-driven two-wheeled vehicle used in cycling. The bicycle is an energy-efficient, nonpolluting form of transport, and it is estimated that 800 million bicycles are in use throughout the world – outnumbering cars three to one. The energy and materials required to make one car are sufficient to produce 100 bicycles.

China, India, Denmark, and the Netherlands are countries with a high use of bicycles. More than 10% of road spending in the Netherlands is on cycleways and bicycle parking and cycling accounts for 50% of urban travel in some Dutch cities. In the UK, although one in three households own a bicycle, average use is only about 6 km/4 mi per week per household. Since 1950, with the growth in ownership and use of motor vehicles bicycle use has declined by about 75% and now only accounts for some 1% of all traffic.

biodegradable capable of being broken down by living organisms, principally bacteria and fungi. In biodegradable substances, such as food and sewage, the natural processes of decay lead to compaction and liquefaction, and to the release of nutrients that are then recycled by the ecosystem. This process can have some disadvantageous side effects, such as the release of methane, an explosive greenhouse gas. However, the technology now exists for waste tips to collect methane in underground pipes, drawing it off and using it as a cheap source of energy. Nonbiodegradable substances, such as glass, heavy metals, and most types of plastic, present serious problems of disposal.

biodiversity (contraction of *biological diversity*) measure of the variety of the Earth's animal, plant, and microbial species; of genetic differences within species; and of the ecosystems that support those species. Its maintenance is important for ecological stability and as a resource for research into, for example, new drugs and crops. Research suggests that biodiversity is far greater than previously realized, especially among smaller organisms – for instance, it is thought that there are 30–40 million insects, of which only a million have so far been identified. It is estimated that 7% of the Earth's surface hosts between 50–75% of the world's biological diversity. The tropics are the richest zones of biodiversity in the world. Costa Rica, for example, has an area less than 10% of the size of France but possesses three times as many vertebrate species.

Today, the most significant threat to biodiversity comes from the destruction of rainforests and other habitats in the southern hemisphere. Such destruction throughout the 20th century is believed to have resulted in the most severe and rapid loss of diversity in the history of the planet.

The term came to public prominence 1992 when an international convention for the preservation of biodiversity was signed by over 100 world leaders at the Earth Summit in Brazil. The convention called on industrialized countries to give financial and technological help to developing countries to allow them to protect and manage their natural resources, and profit from growing commercial demand for genes and chemicals from wild species. However, the convention was weakened by the USA's refusal to sign because of fears it would undermine the patents and licences of biotechnology companies.

biofuel any solid, liquid, or gaseous fuel produced from organic (once living) matter, either directly from plants or indirectly from industrial, commercial, domestic, or agricultural wastes. There are three main

methods for the development of biofuels: the burning of dry organic
wastes (such as household refuse, industrial and agricultural wastes,
straw, wood, and peat); the fermentation of wet wastes (such as animal
dung) in the absence of oxygen to produce biogas (containing up to
60% methane), or the fermentation of sugar cane or corn to produce
alcohol; and energy forestry (producing fast-growing wood for fuel).
Biofuel can be considered a sustainable (although not necessarily
⇨renewable) form of energy as the carbon dioxide produced when it is
burnt is balanced by the carbon dioxide taken out of the atmosphere
when the organic matter was alive and growing.

biogeography the study of how and why plants and animals are distributed around the world, in the past as well as in the present. More
specifically, a theory describing the geographical distribution of species
developed by Robert MacArthur and US zoologist Edward O Wilson.
The theory argues that for many species, ecological specializations
mean that suitable habitats are found in a disparate pattern. Thus the
ponds in which dragonflies breed are separated by large tracts of land,
and edelweiss adapted to alpine peaks cannot colonize the deep valleys
between.

biological oxygen demand (BOD) the amount of dissolved oxygen
taken up by microorganisms in a sample of water. Since these microorganisms live by decomposing organic matter, and the amount of
oxygen used is proportional to their number and metabolic rate, BOD
can be used as a measure of the extent to which the water is polluted
with organic compounds.

biological shield shield around a nuclear reactor that is intended to
protect personnel from the effects of ⇨radiation. It usually consists of
a thick wall of steel and concrete.

biological weathering form of ⇨weathering caused by the activities of living organisms – for example, the growth of roots or the
burrowing of animals. Tree roots are probably the most significant
agents of biological weathering as they are capable of prising apart
rocks by growing into cracks and joints. The action of plants in break-
ing up rocks in this way is an important part of the process of the
development of ⇨soil.

biomass the total mass of living organisms present in a given area. It may be specified for a particular species – for example, earthworm biomass – or for a general category – for example, herbivore biomass. Estimates also exist for the entire global plant biomass. Measurements of biomass can be used to study interactions between organisms, the stability of those interactions, and variations in population numbers.

Some two-thirds of the world's population cooks and heats water by burning biomass, usually wood. Plant biomass can be a renewable source of energy as replacement supplies can be grown relatively quickly. Fossil fuels however, originally formed from biomass, accumulate so slowly that they cannot be considered renewable. The burning of biomass produces 3.5 million tonnes of carbon in the form of carbon dioxide each year, accounting for up to 40% of the world's annual carbon dioxide production.

biome broad natural assemblage of plants and animals shaped by common patterns of vegetation and climate. Examples include the tundra biome and the desert biome.

bioreactor sealed vessel where a variety of microbial reactions can take place. The simplest bioreactors involve the slow decay of vegetable or animal waste, with the emission of methane which can be used as fuel. Laboratory bioreactors control pH, acidity, and oxygen content and are used in advanced biotechnological operations, such as the production of antibiotics by genetically-engineered bacteria.

biosphere the narrow zone that supports life on our planet. It is limited to the waters of the Earth, a fraction of its crust, and the lower regions of the atmosphere.

BioSphere 2 (BS2) ecological test project, a 'planet in a bottle', in Arizona, USA. Under a sealed glass dome, several different habitats were recreated, with representatives of nearly 4,000 species, including eight humans 1991–1993, to see how well air, water, and waste can be recycled in an enclosed environment and whether a stable ecosystem can be created.

The sealed area covered a total of 3.5 acres and contained tropical rainforest, salt marsh, desert, coral reef, and savanna habitats, as well as a section for intensive agriculture. The people within were entirely self-sufficient, except for electricity, which was supplied by a 3.7 megawatt power station on the outside (solar panels were considered too expensive). Experiments with biospheres that contain relatively simple life forms have been carried out for decades, and a 21-day trial period 1989 that included humans preceded the construction of BS2. However, BS2 is not in fact the second in a series: the Earth is considered to be Bio-Sphere 1.

The cost of setting up and maintaining the project has been estimated at $100 million, some of which will be covered by paying visitors, who can view the inhabitants through the geodesic glass dome. The major problem was in maintaining satisfactory oxygen levels, which in late 1993 became so erratic that additional oxygen had to be pumped in.

biotic factor organic variable affecting an ecosystem – for example, the changing population of elephants and its effect on the African savanna.

bird of paradise one of 40 species of crowlike birds, family *Paradiseidae*, native to New Guinea and neighbouring islands. Females are generally drably coloured, but the males have bright and elaborate plumage used in courtship displays. Hunted almost to extinction for their plumage, they are now subject to conservation measures.

birth control another name for ⇨family planning; see also ⇨contraceptive.

birth rate the number of live births per 1,000 of the population of an area over the period of a year. Birth rate is a factor in demographic transition. It is sometimes called *crude birth rate* because it takes in the whole population, including men, and women who are too old to bear children.

In the 20th century, the UK's birth rate has fallen from 28 per 1,000 to less than 10 per 1,000 owing to increased use of contraception, better living standards, and falling infant mortality. The birth rate is declining in most Western countries but remains high in developing countries, adding to the already significant problems of poverty and environmental destruction facing these nations.

bison large, hoofed mammal of the bovine family. There are two species, both brown. The *European bison* or **wisent** *Bison bonasus*, of which only a few protected herds survive, is about 2 m/7 ft high and weighs up to 1,100 kg/2,500 lb. The **North American bison** (often known as 'buffalo') *Bison bison* is slightly smaller, with a heavier mane and more sloping hindquarters. Formerly roaming the prairies in vast numbers, it was almost exterminated in the 19th century, but survives in protected areas.

Buffalos have been crossed with domestic cattle to produce a hardy hybrid, the 'beefalo', which yields a lean carcass on an economical grass diet.

bitumen impure mixture of hydrocarbons, including such deposits as petroleum, asphalt, and natural gas, although sometimes the term is restricted to a soft kind of pitch resembling asphalt.

Solid bitumen may have arisen as a residue from the evaporation of petroleum. If evaporation took place from a pool or lake of petroleum, the residue might form a pitch or asphalt lake, such as Pitch Lake in Trinidad. Bitumen was used in ancient times as a mortar, and by the Egyptians for embalming.

blue-green algae or *cyanobacteria* single-celled, primitive organisms that resemble bacteria in their internal cell organization, sometimes joined together in colonies or filaments. Blue-green algae are among the oldest known living organisms and, with bacteria, belong to the kingdom Monera; remains have been found in rocks up to 3.5 billion years old. They are widely distributed in aquatic habitats, on the damp surfaces of rocks and trees, and in the soil.

Blue-green algae, like bacteria, are prokaryotic organisms. Some can fix nitrogen and thus are necessary to the nitrogen cycle, while others follow a symbiotic existence – for example, living in association with fungi to form lichens. It is believed that the first photosynthetic blue-green algae, living on the primitive Earth, were responsible for enriching the early atmosphere with free oxygen and therefore made it possible for ⇨aerobic organisms to evolve.

Blueprint for Survival environmental manifesto published 1972 in the UK by the editors of the *Ecologist* magazine. The statement of support it attracted from a wide range of scientists helped draw attention to the magnitude of environmental problems.

bog type of wetland where decomposition is slowed down and dead plant matter accumulates as ⇨peat. Bogs develop under conditions of low temperature, high acidity, low nutrient supply, stagnant water, and oxygen deficiency. Typical bog plants are sphagnum moss, rushes, and cotton grass; insectivorous plants such as sundews and bladderworts are common in bogs (insect prey make up for the lack of nutrients). Bogs resist conventional farming methods such as forestry or grazing, but are prized as habitats for wildlife. Draining a bog effectively destroys it.

Botswana *the Okavango swamp is threatened by plans to develop the area for mining and agriculture which will only serve to heighten the country's shortage of water, Botswana's scarcest resource.*

---

bovine spongiform encephalopathy (BSE) disease of cattle, allied to scrapie, that renders the brain spongy and may drive an animal mad. It has been identified only in the UK, where more than 26,000 cases were confirmed between the first diagnosis Nov 1986 and April 1991. However, in 1991 ostriches in a German zoo were found to have the spongy holes in the brain which characterize the condition. The organism causing it is unknown; it is not a conventional virus because it is more resistant to chemicals and heat, cannot be seen even under an electron microscope, cannot be grown in tissue culture, and does not appear to provoke an immune response in the body. BSE is very similar to, and may be related to, Creutzfeldt-Jakob disease and kuru, which affect humans.

The source of the disease has been traced to manufactured protein feed incorporating the rendered brains of scrapie-infected sheep. Following the animal-protein food ban in 1988, there was a single new case of BSE, indicating that the disease could be transmitted from cows to their calves.

Brandt Commission international committee 1977–83 set up to study global development issues. The commission produced two reports, which stress that the countries of the wealthy, industrialized North and the poor South (or developing world) are dependent on one another. The reports suggested ways by which resources could be transferred to the Southern nations, together with measures that could be taken by the South to reduce poverty and increase food production.

Brazil *Brazil has one-third of the world's tropical rainforest. It contains 55,000 species of flowering plants (the greatest variety in the world) and 20% of all the world's bird species. During the 1980s at least 8% of the Amazon rainforest, amounting to 404,000 sq km/ 156,000 sq mi, was destroyed by settlers who cleared the land for cultivation and grazing. In urban areas, overpopulation has resulted in air and water pollution. In particular, the dumping of untreated waste has contaminated many waterways.*

breeder reactor or *fast breeder* alternative names for ⇨fast reactor, a type of nuclear reactor.

breeding in biology, the crossing and selection of animals and plants to change the characteristics of an existing breed or to produce a new one (see ⇨artificial selection).

breeding in nuclear physics, a process in a reactor in which more fissionable material is produced than is consumed in running the reactor.

  For example, plutonium-239 can be made from the relatively plentiful (but nonfissile) uranium-238, or uranium-233 can be produced from thorium. The Pu-239 or U-233 can then be used to fuel other reactors. The French breeder reactor Superphénix, one of the most successful, generates 250 megawatts of electrical power.

bromine dark, reddish-brown, nonmetallic element, a volatile liquid at room temperature, symbol Br, atomic number 35, relative atomic mass 79.904. It is a member of the halogen group, has an unpleasant odour, and is very irritating to mucous membranes. Its salts are bromides.

  Bromine was formerly extracted from salt beds but is now mostly obtained from sea water, where it occurs in small quantities. Its compounds are used in photography and in the chemical and pharmaceutical industries.

BSE abbreviation for ⇨*bovine spongiform encephalopathy*.

Bulgaria *pollution has virtually eliminated all species of fish once caught in the Black Sea. Vehicle-exhaust and industrial emissions in*

*Sofia and other major urban centres have led to dust concentrations
more than twice the medically accepted level. Large areas of land in
rural areas have been contaminated by heavy metals and suffered soil
degradation due to mining operations and dumping of waste.*

bushman's rabbit or *riverine rabbit* a wild rodent *Bunolagus monticulari* found in dense riverine bush in South Africa. It lives in small populations, and individuals are only seen very occasionally; it is now at extreme risk of extinction owing to loss of habitat to agriculture.

bustard bird of the family Otididae, related to cranes but with a rounder body, a thicker neck, and a relatively short beak. Bustards are found on the ground on open plains and fields.

They are found in N Asia and Europe, although there are fewer than 30,000 great bustards left in Europe and two-thirds of these live on the Spanish steppes. The bustard has been extinct in Britain for some time, although attempts are being made by the Great Bustard Trust to naturalize it again on Salisbury Plain. The great Indian bustard is endangered because of hunting and loss of its habitat to agriculture; there are less than 1,000 individuals left.

butterfly insect belonging, like moths, to the order Lepidoptera, in which the wings are covered with tiny scales, often brightly coloured. There are some 15,000 species of butterfly, many of which are under threat throughout the world because of the destruction of habitat.

by-product substance formed incidentally during the manufacture of some other substance; for example, slag is a by-product of the production of iron in a blast furnace. For industrial processes to be economical, by-products must be recycled or used in other ways as far as possible; in this example, slag is used for making roads.

Often, a poisonous by-product is removed by transforming it into another substance, which although less harmful is often still inconvenient. For example, the sulphur dioxide produced as a by-product of electricity generation can be removed from the smoke stack using ⇨flue gas desulphurization. This process produces gypsum, some of which can be used in the building industry.

C

cadmium soft, silver-white, ductile, and malleable metallic element, symbol Cd, atomic number 48, relative atomic mass 112.40. Cadmium occurs in nature as a sulphide or carbonate in zinc ores. It is a toxic metal that, because of industrial dumping, has become an environmental pollutant. It is used in batteries, electroplating, and as a constituent of alloys used for bearings with low coefficients of friction; it is also a constituent of an alloy with a very low melting point. Cadmium is used in the control rods of nuclear reactors, because of its high absorption of neutrons.

caesium soft, silvery-white, ductile metallic element, symbol Cs, atomic number 55, relative atomic mass 132.905. It is one of the alkali metals, and is the most electropositive of all the elements. In air it ignites spontaneously, and it reacts vigorously with water. It is used in the manufacture of photoelectric cells. The name comes from the blueness of its spectral line.

The rate of vibration of caesium atoms is used as the standard of measuring time. Its radioactive isotope Cs-137 (half-life 30.17 years) is a product of fission in nuclear explosions and in nuclear reactors; it is one of the most dangerous waste products of the nuclear industry, being a highly radioactive biological analogue for potassium.

Cameroon *logging and poaching pose major threats to forest areas and wildlife habitat across the country, even in reserves. The Korup National Park preserves 1,300 sq km/500 sq mi of Africa's fast-disappearing tropical rainforest and scientists have identified nearly 100 potentially useful chemical substances produced naturally by the plants of this forest.*

Canada acid rain from industrial pollution in Canada and the US is a significant problem; sugar maples are dying in E Canada as a result of increasing soil acidification and nine rivers in Nova Scotia are now too acid to support salmon or trout reproduction. Logging is also a major issue as environmentalists mount increasingly vociferous campaigns to save the country's ancient woodland and loggers defend their right to exploit the resource on which their livelihood depends.

cancer group of diseases characterized by abnormal proliferation of cells. Cancer (malignant) cells are usually degenerate, capable only of reproducing themselves (tumour formation). Malignant cells tend to spread from their site of origin by travelling through the bloodstream or lymphatic system.

There are more than 100 types of cancer. Some, like lung or bowel cancer, are common; others are rare. The likely cause remains unexplained. Triggering agents (⇨carcinogens) include chemicals such as those found in cigarette smoke, other forms of smoke, asbestos dust, exhaust fumes, and many industrial chemicals. Some viruses can also trigger the cancerous growth of cells, as can X-rays and radioactivity. Dietary factors are important in some cancers; for example, lack of fibre in the diet may predispose people to bowel cancer and a diet high in animal fats and low in fresh vegetables and fruit increases the risk of breast cancer. Psychological stress may increase the risk of cancer, more so if the person concerned is not able to control the source of the stress. In some families there is a genetic tendency toward a particular type of cancer.

Cancer is one of the leading causes of death in the industrialized world, yet it is by no means incurable. Overall, however, the total occurrence of cancer appears to be on the increase. Increasingly, cancer prevention, including monitoring environmental factors, is coming to dominate cancer research.

capuchin monkey of the genus *Cebus* found in Central and South America. They live in small groups, feed on fruit and insects, and have long tails that are semi-prehensile and can give support when climbing through the trees.

There are now thought to be only 800 yellow-breasted capuchins left in the wild, found only in the Atlantic forest in S of Bahia state.

car small, driver-guided, passenger-carrying motor vehicle; originally the automated version of the horse-drawn carriage, meant to convey people and their goods over streets and roads. Over 50 million motor cars are produced each year worldwide. Most are four-wheeled and have water-cooled, piston-type ⇨internal-combustion engines fuelled by petrol or diesel, causing many pollution problems. Variations have existed for decades that use ingenious and often nonpolluting power plants, but the motor industry long ago settled on this general formula for the consumer market.

During the 1980s the number of cars in the UK increased by 34% to some 22 million, whereas in 1950 there were only 4 million. In the UK, 85% of all passenger transport is by car. The convenience of and degree of independence afforded to individuals by cars should be balanced against their harmful effects. Cars are responsible for 85% of all UK carbon monoxide pollution and 45% of nitrogen oxide pollution. In the presence of sunlight, car exhaust fumes react to produce ground-level ozone, an important constituent of photochemical ⇨smog.

carbon cycle sequence by which carbon circulates and is recycled through the natural world. The carbon element from carbon dioxide, released into the atmosphere by living things as a result of ⇨respiration, is taken up by plants during ⇨photosynthesis and converted into carbohydrates; the oxygen component is released back into the atmosphere. The simplest link in the carbon cycle occurs when an animal eats a plant and carbon is transferred from, say, a leaf cell to the animal body. Today, the carbon cycle is in danger of being disrupted by the increased consumption and burning of fossil fuels, and the burning of large tracts of tropical forests, as a result of which levels of carbon dioxide are building up in the atmosphere and probably contributing to the ⇨greenhouse effect.

The European Community has pledged to stabilize CO_2 emissions at 1990 levels by the year 2000. They would otherwise rise by 11% per year over this period. It is hoped to maintain such levels by burning less fossil fuel through increased energy efficiency.

carbon dioxide CO_2 colourless, odourless gas, slightly soluble in

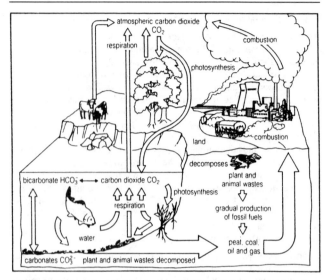

carbon cycle The carbon cycle is necessary for the continuation of life. Since there is only a limited amount of carbon in the Earth and its atmosphere, carbon must be continuously recycled if life is to continue. Other chemicals necessary for life – nitrogen, sulphur, and phosphorus, for example – also circulate in natural cycles.

water and denser than air. It is formed by the complete oxidation of carbon. It is produced by living things during the processes of respiration and the decay of organic matter, and plays a vital role in the ⇨carbon cycle. It is used as a coolant in its solid form (known as 'dry ice'), and in the chemical industry. Its increasing density contributes to the ⇨greenhouse effect and ⇨global warming. Britain has 1% of the world's population, yet it produces 3% of CO_2 emissions; the USA has 5% of the world's population and produces 25% of CO_2 emissions.

Carbon dioxide is vital to life on Earth, not only because it is an ingredient in plant ⇨photosynthesis, but also because it raises the atmospheric temperature through the greenhouse effect. Without the presence of CO_2, the planet would be many degrees cooler. However, the increase in levels of CO_2 in the 20th century threaten to accelerate the greenhouse effect to an unacceptable degree, causing global warming and dramatic climate change.

carbon monoxide CO colourless, odourless gas formed when carbon is oxidized in a limited supply of air. It is a poisonous constituent of car exhaust fumes, forming a stable compound with haemoglobin in the blood, thus preventing the haemoglobin from transporting oxygen to the body tissues.

carcinogen any agent that increases the chance of a cell becoming cancerous (see ⇨cancer), including various chemical compounds, some viruses, X-rays, and other forms of ionizing radiation. The term is often used more narrowly to mean chemical carcinogens only.

carrying capacity the maximum number of animals of a given species that a particular area can support. When the carrying capacity is exceeded, there is insufficient food (or other resources) for the members of the population. The population may then be reduced by emigration, reproductive failure, or death through starvation.

cash crop crop grown solely for sale rather than for the farmer's own use, for example, coffee, cotton, or sugar beet. Many Third World countries grow cash crops to meet their debt repayments rather than grow food for their own people, thus adding to the problems of malnutrition and starvation. As cash crops are grown intensively, usually as a monoculture, environmental damage by pesticides and fertilizers is often unavoidable. The price for these crops depends on financial interests, such as those of the multinational companies and the International Monetary Fund.

In 1990 Uganda, Rwanda, Nicaragua, and Somalia were the countries most dependent on cash crops for income. In Britain, the most widespread cash crop is the potato.

catalytic converter device fitted to the exhaust system of a motor vehicle in order to reduce toxic emissions from the engine. It converts harmful exhaust products to relatively harmless ones by passing the exhaust gases over a mixture of catalysts coated on a metal or ceramic honeycomb, a structure that increases the surface area and therefore the amount of active catalyst with which the exhaust gases will come into contact. *Oxidation catalysts* (small amounts of precious palladium and platinum metals) convert hydrocarbons (unburnt fuel) and carbon monoxide into carbon dioxide and water, but do not affect nitrogen oxide emissions. *Three-way catalysts* (platinum and rhodium metals) convert nitrogen oxide gases into nitrogen and oxygen as well as treating carbon monoxide and hydrocarbons. Catalytic converters are destroyed by emissions from leaded petrol and work best at a temperature of 300°C/570°F. The benefits of catalytic converters are offset by any increase in the number of cars in use.

Over the lifetime of a vehicle, a catalytic converter can reduce hydrocarbon emissions by 87%, carbon monoxide emissions by 85%, and nitrogen oxide emissions by 62%, but will cause a slight increase in the amount of carbon dioxide emitted. Catalytic converters are standard in the USA, where a 90% reduction in pollution from cars was achieved without loss of engine performance or fuel economy.

catastrophism theory that the geological features of the Earth were formed by a series of sudden, violent 'catastrophes' beyond the ordinary workings of nature. The theory was largely the work of Georges Cuvier (1769–1832) and has now been discredited.

cellulose complex carbohydrate composed of long chains of glucose units. It is the principal constituent of the cell wall of higher plants, and a vital ingredient in the diet of many herbivores. Molecules of cellulose are organized into long, unbranched microfibrils that give support to the cell wall. No mammal produces the enzyme (cellulase) necessary for digesting cellulose; mammals such as rabbits and cows are only able to digest grass because the bacteria present in their gut manufacture the appropriate enzyme.

Central African Republic poaching for ivory has decimated the elephant population to less than 10% of levels even 40 years ago. However, conservation measures such as a ban on hunting 1985 and the creation of reserves are starting to reverse the decline. An estimated 87% of the urban population has no access to safe drinking water.

CERN nuclear research organization founded 1954 as a cooperative enterprise among European governments. It has laboratories at Meyrin, near Geneva, Switzerland. It was originally known as the *Conseil Européen pour la Recherche Nucléaire* but subsequently renamed *Organisation Européenne pour la Recherche Nucléaire*, although still familiarly known as CERN. It houses the world's largest particle accelerator, the ⇨Large Electron–Positron Collider (LEP), with which notable advances have been made in particle physics.

CFC abbreviation for ⇨chlorofluorocarbon.

chain reaction mechanism that produces very fast, exothermic reactions, as in the formation of flames and explosions.

The reaction begins with the formation of a single reactive molecule. This combines with an inactive molecule to form two reactive molecules. These two produce four (or more) reactive molecules; very quickly, very many reactive molecules are produced, so the reaction rate accelerates dramatically. The reactive molecules contain an unpaired electron and are called ⇨free radicals; they last only a short time because they are so reactive.

cheetah large wild cat *Acinonyx jubatus* native to Africa, Arabia, and SW Asia, but now rare in some areas. Yellowish with black spots, it has a slim lithe build. It is up to 1 m/3 ft tall at the shoulder, and up to 1.5 m/5 ft long. It can reach 110 kph/70 mph making it the world's fastest mammal. Cheetahs sustain this speed during short chases only – they tire after about 400 m. Cheetahs live in open country where they hunt small antelopes, hares, and birds.

Cheetahs face threats both from ranchers who shoot them as vermin and from general habitat destruction that is reducing the prey on which they feed, especially gazelles. As a result the wild population is thought to have fallen by over half since the 1970s; there are now thought to be no more than 5,000–12,0000 left.

chelate chemical compound whose molecules consist of one or more metal atoms or charged ions joined to chains of organic residues by coordinate (or dative covalent) chemical bonds.

The parent organic compound is known as a *chelating agent* – for example, EDTA (ethylene-diaminetetraacetic acid), used in chemical analysis. Chelates are used in analytical chemistry, in agriculture and horticulture as carriers of essential trace metals, in water softening, and in the treatment of thalassaemia by removing excess iron, which may build up to toxic levels in the body. Metalloproteins (natural chelates) may influence the performance of enzymes or provide a mechanism for the storage of iron in the spleen and plasma of the human body.

chemical oxygen demand (COD) measure of water and effluent quality, expressed as the amount of oxygen (in parts per million) required to oxidize the reducing substances present.

Under controlled conditions of time and temperature, a chemical oxidizing agent (potassium permanganate or dichromate) is added to the sample of water or effluent under consideration, and the amount needed to oxidize the reducing materials present is measured. From this the chemically equivalent amount of oxygen can be calculated. Since the reducing substances typically include remains of living organisms, COD may be regarded as reflecting the extent to which the sample is polluted. Compare ⇨biological oxygen demand.

chemical weathering form of weathering brought about by a chemical change in the rocks affected. Chemical weathering involves the 'rotting', or breakdown, of the minerals within a rock, and usually produces a claylike residue (such as china clay and bauxite). Some chemicals are dissolved and carried away from the weathering source.

A number of processes bring about chemical weathering, such as carbonation (breakdown by weakly acidic rainwater), ⇨hydrolysis (breakdown by water), hydration (breakdown by the absorption of water), and oxidation (breakdown by the oxygen in water).

Chernobyl town in central Ukraine; site of a nuclear power station. In April 1986, two huge explosions destroyed a central reactor, breaching the 1,000 tonne roof. In the immediate vicinity of Chernobyl, 31 people died and 135,000 were permanently evacuated. It has been estimated that there will be an additional 20–40,000 deaths from cancer in the next 60 years.

The resulting clouds of radioactive isotopes were traced all over Europe, from Ireland to Greece. Together with the fallout from nuclear weapons testing conducted in the past, the Chernobyl explosion currently accounts for half of the ⇨background radiation in the UK, with the greatest effects on average concentration occurring in Scotland and Northern Ireland.

Current attempts to make the land around Chernobyl safe to farm again include scraping off the top 3–4 cm/1–1.5 in of topsoil and then burying this 45 cm/18 in below without disturbing the intervening layer.

China air pollution is a major concern, particularly in large urban concentrations, where pollution is often so bad as to threaten health. More cities in China exceed the World Health Organization's recommended safe limits for air pollution than in any other country. Deforestation and loss of wildlife habitat are severe in many areas of the country as a result of the drives for economic growth, particularly in steel production, of the Mao years.

---

chinchilla South American rodent *Chinchilla laniger* found in high, rather barren areas of the Andes in Bolivia and Chile. About the size of a small rabbit, it has long ears and a long bushy tail, and shelters in rock crevices. These gregarious animals have thick, soft, silver-grey fur, and were hunted almost to extinction for it. They are now farmed and protected in the wild.

Chipko movement Indian grass-roots villagers' movement campaigning against the destruction of their forests. Its broad principles are nonviolent direct action, a commitment to the links between village life and an unplundered environment, and a respect for all living things.

The Chipko movement originated in the Indian Himalayas during the 1970s but still lives on in India and elsewhere, inspired by the teachings of Mahatma Gandhi. The movement was named after the Hindi word for 'to embrace', from the villagers' original tactics of embracing trees to prevent them being felled.

chlordane organochlorine pesticide, used especially against ants and termites.

chlorination the treatment of water with chlorine in order to disinfect it; also, any chemical reaction in which a chlorine atom is introduced into a chemical compound.

chlorine greenish-yellow, gaseous, nonmetallic element with a pungent odour, symbol Cl, atomic number 17, relative atomic mass 35.453 (a mixture of two isotopes: 75% Cl-35, 25% Cl-37). It is the second member of the halogen group (group VII of the periodic table). In nature it is widely distributed in combination with the alkali metals, as chlorates or chlorides; in its pure form the gas is a diatomic molecule (Cl_2). It is a very reactive element, and combines with most metals, some nonmetals, and a wide variety of compounds.

Industrially, chlorine is prepared by the electrolysis of concentrated brine. It is used in making bleaches, in sterilizing water for drinking and for swimming baths, and in the manufacture of chloro-organic compounds such as chlorinated solvents, CFCs, and PVC. During World War I it was used in gas warfare; it sears the membranes of the nose, throat, and lungs, producing pneumonia.

chlorofluorocarbon (CFC) synthetic chemical that is odourless, nontoxic, nonflammable, and chemically inert. The first CFC was synthesized 1892, but no use was found for it until the 1920s. Since then their stability and apparently harmless properties have made CFCs popular as propellants in ⇨aerosol cans, as refrigerants in refrigerators and air conditioners, and in the manufacture of foam packaging. However, they are now known to destroy ⇨ozone in the upper atmosphere, the zone which filters harmful UV-B radiation. In June 1990 representatives of 93 nations, including the UK and the USA, agreed to phase out production of CFCs and various other ozone-depleting chemicals by the end of the 20th century.

When CFCs are released into the atmosphere, they drift up slowly into the stratosphere, where, under the influence of ultraviolet radiation from the Sun, they break down into chlorine atoms which destroy the ozone layer and allow harmful radiation from the Sun to reach the Earth's surface. CFCs can remain in the atmosphere for more than 100 years. Replacements for CFCs are being developed, and research into safe methods of destroying existing CFCs is being carried out. The European Community has agreed to ban the five 'full hydrogenated' CFCs that are restricted under the ⇨Montréal Protocol and a range of CFCs used as industrial solvents, refrigerants, and in fire extinguishers by the end of 1995.

chlorosis abnormal condition of green plants in which the stems and leaves turn pale green or yellow. The yellowing is due to a reduction in the levels of the green chlorophyll pigments. It may be caused by a deficiency in essential elements (such as magnesium, iron, or manganese which may be lacking because of ⇨leaching), a lack of light, genetic factors, or viral infection.

CITES (abbreviation for *Convention on International Trade in Endangered Species*) international agreement under the auspices of the ⇨IUCN with the aim of regulating trade in ⇨endangered species of animals and plants. The agreement came into force 1975 and by 1991 had been signed by 110 states. It prohibits any trade in a category of 8,000 highly endangered species and controls trade in a further 30,000 species.

The international trade in animal and plants is worth some 3 billion a year, one third of which is illegal. This trade is a central cause of some species – such as the tiger and the rhinoceros – being driven to the verge of extinction.

Clean Air Act legislation designed to improve the quality of air by enforcing pollution controls on industry and households. The first Clean Air Act in the UK was passed 1956 after the London Smog killed 4,000 people. The USA enacted a Clean Air Act 1970, further amended 1990.

In Europe, national legislation is supplemented by EC and UN agreements and conventions such as EC regulations on levels of ozone, nitrogen oxide, and sulphur dioxide. In addition the UK has signed protocols drawn up by the UN Economic Commission for Europe as part of an attempt to reduce transboundary pollution.

The US Clean Air Act has been partially successful; as in Europe, levels of sulphur dioxide have been stabilized or reduced, but other pollutants, including the nitrogen oxides, are increasing. At least 86 million US citizens live in areas where air standards are violated, by pollutants such as sulphur dioxide, the nitrogen oxides, black smoke, carbon monoxide, and airborne lead. Progress by legislation was slow so a new market-led policy based on the ⇨polluter-pays principle was introduced in the 1990 Amendments to the Clean Air Act. Under this

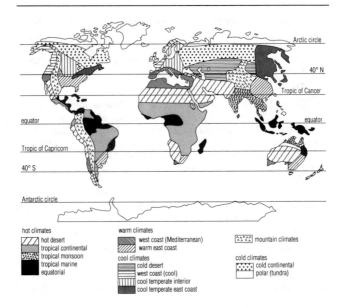

hot climates
- hot desert
- tropical continental
- tropical monsoon
- tropical marine
- equatorial

warm climates
- west coast (Mediterranean)
- warm east coast

cool climates
- cold desert
- west coast (cool)
- cool temperate interior
- cool temperate east coast

mountain climates

cold climates
- cold continental
- polar (tundra)

The world's climatic zones. There are many sytems of classifying climate. One system, that of Wladimir Köppen, was based on temperature and plant type. Other systems take into account the distribution of global winds.

policy, companies which wish to pollute must buy vouchers allowing them to do so. The vouchers are deliberately expensive so that it is generally cheaper for companies to install anti-pollution technology.

climate weather conditions at a particular place over a period of time. Climate encompasses all the meteorological elements and the factors that influence them. The primary factors that determine the variations of climate over the surface of the Earth are: (a) the effect of latitude and the tilt of the Earth's axis to the plane of the orbit about the Sun (66.5°);

(b) the large-scale movements of different wind belts over the Earth's surface; (c) the temperature difference between land and sea; (d) contours of the ground; and (e) location of the area in relation to ocean currents. Catastrophic variations to climate may be caused by the impact of another planetary body, or by clouds resulting from volcanic activity. Recent research indicates that human activity may influence world climate as average temperatures rise (see ⇨global warming). However, because of the extremely complex nature of climatic systems, it is very difficult to specify exactly what changes will result from global warming. The most important local or global meteorological changes brought about by human activity are those linked with ⇨ozone depleters and the ⇨greenhouse effect.

How much heat the Earth receives from the Sun varies in different latitudes and at different times of the year. In the equatorial region the mean daily temperature of the air near the ground has no large seasonal variation. In the polar regions the temperature in the long winter, when there is no incoming solar radiation, falls far below the summer value. Climate types were first classified by Vladimir Köppen (1846–1940) in 1884. The temperature of the sea, and of the air above it, varies little in the course of day or night, whereas the surface of the land is rapidly cooled by lack of solar radiation. In the same way the annual change of temperature is relatively small over the sea and great over the land. Continental areas are thus colder than the sea in winter and warmer in summer. Winds that blow from the sea are warm in winter and cool in summer, while those winds from the central parts of continents are hot in summer and cold in winter. On average, air temperature drops with increasing land height at a rate of 1°C/1.8°F per 90 m/300 ft. Thus places situated above mean sea level usually have lower temperatures than places at or near sea level. Even in equatorial regions, high mountains are snow-covered during the whole year.

Rainfall is produced by the condensation of water vapour in air. When winds blow against a range of mountains so that the air is forced to ascend, rain results, the amount depending on the height of the ground and the dampness of the air. The complexity of the distribution of land and sea, and the consequent complexity of the general circulation of the atmosphere, have a direct effect on the distribution of the climate. Centred on the equator is a belt of tropical rainforest, which may be either constantly wet or monsoonal (seasonal with wet and dry seasons in each year). On each side of this is a belt of savanna, with lighter seasonal rainfall and less dense vegetation, largely in the form of grasses. Usually there is then a transition through steppe (semi-arid) to desert (arid), with a further transition through steppe to Mediterranean climate with dry summer, followed by the moist temperate climate of middle latitudes. Next comes a zone of cold climate with moist winter. Where the desert extends into middle latitudes, however, the zones of Mediterranean and moist temperate climates are missing, and the transition is from desert to a cold climate with moist winter. In the extreme east of Asia a cold climate with dry winters extends from about 70° N to 35° N. The polar caps have ⇨tundra and glacial climates, with little or no precipitation (rain or snow).

climax community assemblage of plants and animals that is relatively stable in its environment. It is brought about by ecological ⇨succession, and represents the point at which succession ceases to occur.

In temperate or tropical conditions, a typical climax community comprises woodland or forest and its associated fauna (for example, an oak wood in the UK). In essence, most land management is a series of interferences with the process of succession.

climax vegetation the plants in a climax community.

coal black or blackish mineral substance formed from the compaction of ancient plant matter in tropical swamp conditions. It is used as a fuel and in the chemical industry. Coal is classified according to the proportion of carbon it contains. The main types are ⇨anthracite (shiny, with more than 90% carbon), *bituminous* (shiny and dull patches, more than 80% carbon), and *lignite* (woody, grading into peat, 70% carbon).

In the second half of the 18th century, coal fuelled the Industrial Revolution. From about 1800, it was used to produce coalgas for gas lighting, and coke for smelting iron ore. More recently it has been used by the petrochemical industry in the production of plastics. Coal burning is one of the main causes of ⇨acid rain. Reductions in UK sulphur dioxide emissions, apparent since 1988, are due mainly to reductions in

eroded headland at low tide

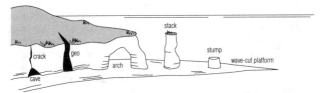

coastal erosion

the burning of coal, as many power stations are now fuelled by gas or imported coal with a lower sulphur content.

coal mining extraction of coal from the Earth's crust. Coal mines may be opencast (see ⇨opencast mining), adit, or deepcast. The least expensive is opencast but this results in scars on the landscape.

coastal erosion the erosion of the land by the constant battering of waves. The force of the waves creates a hydraulic effect compressing air to form explosive pockets in the rocks and cliffs. Rocks and pebbles may be flung against the cliff face (the process of corrasion) and wear it away. Cliffs of chalk and limestone may be dissolved by the process of ⇨solution.

Where resistant rocks form headlands, the sea erodes the coast in successive stages. First it creates cracks in cave openings and then gradually wears away the interior of the caves until their roofs are pierced through to form blowholes. In time, caves at either side of a headland may unite to form a natural arch. When the roof of the arch collapses, a stack is formed. This may be worn down further to produce a stump and a wave-cut platform.

coastal protection measures taken to prevent ⇨coastal erosion. Many stretches of coastline are so severely affected by erosion that beaches are swept away, threatening the livelihood of seaside resorts, and buildings become unsafe, as when a hotel in Scarborough, N Yorkshire, collapsed into the sea 1993.

To reduce erosion, several different forms of coastal protection may be employed. Structures such as sea walls attempt to prevent waves reaching the cliffs by deflecting them back to sea. Such structures are expensive and of limited success. A currently preferred option is to add sediment (beach nourishment) to make a beach wider. This causes waves to break early so that they have less power when they reach the cliffs. Groynes may also be constructed to trap sediment and widen beaches. Increasingly, some areas are being allowed to erode completely in order to concentrate protective measures in other areas.

cocktail effect the effect of two toxic, or potentially toxic, chemicals when taken together rather than separately. Such effects are known to occur with some mixtures of chemicals, with one ingredient making the body more sensitive to another ingredient. This sometimes occurs because both chemicals require the same enzyme to break them down. Chemicals such as pesticides and food additives are only ever tested singly, not in combination with other chemicals that may be consumed at the same time, so no allowance is made for cocktail effects.

COD abbreviation for ⇨*chemical oxygen demand*, a measure of water and effluent quality.

coke clean, light fuel produced by the carbonization of certain types of coal. When this coal is strongly heated in airtight ovens (in order to release all volatile constituents), the brittle, silver-grey remains are coke. Coke comprises 90% carbon together with very small quantities of water, hydrogen, and oxygen, and makes a useful industrial and domestic fuel.

colonization the spread of species into a new habitat, such as a freshly cleared field, a new motorway verge, or a recently flooded valley. The first species to move in are called *pioneers*, and may establish conditions that allow other animals and plants to move in (for example, by improving the condition of the soil or by providing shade). Over time a range of species arrives and the habitat matures; early colonizers will probably be replaced, so that the variety of animal and plant life present changes. This is known as ⇨succession.

combined heat and power generation (CHP generation) simultaneous production of electricity and useful heat in a power station. The heat is often in the form of hot water or steam, which can be used for local district heating or in industry. The electricity output from a CHP plant is lower than from a conventional station, but the overall efficiency of energy conversion is higher. A typical CHP plant may convert 80% of the original fuel energy into a mix of electricity and useful heat, whereas a conventional power station rarely even manages a 40% conversion rate.

Common Agricultural Policy (CAP) system that allows the member countries of the European Community (EC) jointly to organize and control agricultural production within their boundaries. The objectives of the CAP were outlined in the Treaty of Rome: to increase agricultural productivity, to provide a fair standard of living for farmers and their employees, to stabilize markets, and to assure the availability of supply at a price that was reasonable to the consumer. The CAP is increasingly criticized for its role in creating overproduction, and consequent environmental damage, and for the high price of food subsidies.

common land unenclosed wasteland, forest, and pasture used in common by the community at large. Poor people have throughout history gathered fruit, nuts, wood, reeds, roots, game, and so on from common land; in dry regions of India, for example, the landless derive 20% of their annual income in this way, together with much of their food and fuel.

community an assemblage of plants, animals, and other organisms living within a circumscribed area. Communities are usually named by reference to a dominant feature such as characteristic plant species (for example, beechwood community), or a prominent physical feature (for example, a freshwater-pond community).

competition the interaction between two or more organisms, or groups of organisms (for example, species), that use a common resource which is in short supply. Competition invariably results in a reduction in the numbers of one or both competitors, and in evolution contributes both to the decline of certain species and to the evolution of adaptations.

Thus plants may compete with each other for sunlight, or nutrients from the soil, while animals may compete amongst themselves for food, water, or refuge. Competition between individuals of the same species is termed *intraspecific*, whereas that between individuals of two or more species is termed *interspecific*.

compost organic material decomposed by bacteria under controlled conditions to make a nutrient-rich natural fertilizer for use in gardening or farming. A well-made compost heap reaches a high temperature during the composting process, killing most weed seeds that might be present.

condor large bird, a New World vulture *Vultur gryphus*, with wingspan up to 3 m/10 ft, weight up to 13 kg/28 lb, and length up to 1.2 m/3.8 ft. It is black, with some white on the wings and a white frill at the base of the neck. It lives in the Andes and along the South American coast, and feeds on carrion. The Californian condor *Gymnogyps californianus* is a similar bird, on the verge of extinction as it lays only one egg at a time and may not breed every year. It is the subject of a special conservation effort.

Congo deforestation and illegal hunting threaten the country's biodiversity; the lack of effectively organized and resourced conservation measures make the problem acute. An estimated 93% of the rural population is without access to safe drinking water.

---

coniferous forest forest consisting of evergreen trees such as pines and firs. Most conifers grow quickly and can tolerate poor soil, steep slopes, and short growing seasons. Coniferous forests are widespread in Scandinavia and upland areas of the UK such as the Scottish Highlands, and are often planted in ⇨afforestation schemes. They are highly efficient for wood production but provide an impoverished habitat and can result in acidified soil. Conifers also grow in ⇨woodland.

conservation action taken to protect and preserve the natural world, usually from pollution, overexploitation, and other harmful features of human activity. The late 1980s saw a great increase in public concern for the environment, with membership of conservation groups, such as ⇨Friends of the Earth, the Sierra Club, and the Nature Conservancy rising sharply. Globally the most important issues include the depletion of atmospheric ⇨ozone by the action of ⇨chlorofluorocarbons (CFCs), the build-up of carbon dioxide in the atmosphere (thought to contribute to an intensification of the ⇨greenhouse effect), and ⇨deforestation.

Action by governments has been prompted and supplemented by private agencies, such as the World Wide Fund for Nature. In attempts to save particular species or habitats, a distinction is often made between *preservation*, that is maintaining the pristine state of nature exactly as it was or might have been, and *conservation*, the management of natural

conservation: chronology

1681	The last dodo, a long-standing symbol of the need for species conservation, died on the island of Mauritius.
1948	The International Union for Conservation of Nature and Natural Resources (IUCN) was founded, with its sister organization, the World Wildlife Fund (WWF).
1970	The Man and the Biosphere Programme was initiated by UNESCO, providing for an international network of biosphere reserves.
1971	The Convention on Wetlands of International Importance (especially concerned with wildfowl habitat) signed in Ramsar, Iran started a List of Wetlands of International Importance.
1972	The Convention Concerning the Protection of the World Cultural and Natural Heritage adopted in Paris, France, providing for the designation of World Heritage Sites.
1972	The UN Conference on the Human Environment held in Stockholm, Sweden, leading to the creation of the UN Environment Programme (UNEP).
1973	The Convention on International Trade in Endangered Species of Wild Fauna and Flora (CITES) signed in Washington DC.
1974	The world's largest protected area, the Greenland National Park covering 97 million hectares, created.
1980	The World Conservation Strategy launched by the IUCN, with the WWF and UNEP, showing how conservation contributes to development.
1982	The first herd of 10 Arabian oryx bred from a 'captive breeding' programme released into the wild in Oman. The last wild oryx had been killed 1972.
1986	The first 'Red List' of endangered animal species compiled by IUCN.
1989	International trade in ivory banned under CITES legislation in an effort to protect the African elephant from poachers.
1992	The UN convened the 'Earth Summit' in Rio de Janeiro, Brazil, to discuss global planning for a sustainable future. The Convention on Biological Diversity and the Convention on Climate Change were opened for signing.
1993	The Convention on Biological Diversity came into force.

resources in such a way as to integrate the requirements of the local human population with those of the animals, plants, or the habitat being conserved.

In the UK the conservation debate has centred on road-building schemes, the safety of ⇨nuclear energy, and ⇨animal rights. Conservation groups in the UK originated in the 1860s; they include the Commons Preservation Society 1865, which fought successfully against the enclosure of Hampstead Heath (1865) and Epping Forest (1866) in London, the National Footpaths Preservation Society 1844, and the National Trust 1895.

contraceptive any drug, device, technique, or operation that prevents pregnancy. See also ⇨family planning and ⇨birth rate.

copper orange-pink, very malleable and ductile, metallic element, symbol Cu, atomic number 29, relative atomic mass 63.546. It is used for its durability, pliability, high thermal and electrical conductivity, and resistance to corrosion.

Copper is an essential mineral for most living organisms but the copper ion is dangerous as a pollutant. Discharges of copper as an ⇨effluent from factories and commercial mining operations can have a serious effect on water systems, killing algae and fish even in small concentrations. Copper can also pollute the soil due to mining activities and its use in agriculture. If copper enters the food chain, it can cause brain damage.

coppicing woodland management practice of severe pruning where trees are cut down to near ground level at regular intervals, typically every 3–20 years, to promote the growth of numerous shoots from the base.

This form of ⇨forestry was once commonly practised in Europe, principally on hazel and chestnut, to produce large quantities of thin branches for firewood, fencing, and so on; alder, eucalyptus, maple, poplar, and willow were also coppiced. The resulting thicket was known as a coppice or copse. See also ⇨pollarding. Some forests in the UK, such as Epping Forest near London, have coppice stretching back to the Middle Ages.

coral marine invertebrate of the class Anthozoa in the phylum

Cnidaria, which also includes sea anemones and jellyfish. It has a skeleton of lime (calcium carbonate) extracted from the surrounding water. Corals exist in warm seas, at moderate depths with sufficient light. Some coral is valued for decoration or jewellery, for example, Mediterranean red coral *Corallum rubrum*.

Corals have a relationship to the fish that rest or take refuge within their branches, and which excrete nutrients that make the corals grow faster. The majority of corals form large colonies (although there are species that live singly). Their accumulated skeletons make up large coral reefs and atolls. The Great Barrier Reef, to the NE of Australia, is about 1,600 km/1,000 mi long, has a total area of 20,000 sq km/7,700 sq mi, and adds 50 million tonnes of calcium to the reef each year. The world's reefs cover an estimated 620,000 sq km/240,000 sq mi.

Fringing reefs are so called because they build up on the shores of continents or islands, the living animals mainly occupying the outer edges of the reef. *Barrier reefs* are separated from the shore by a saltwater lagoon, which may be as much as 30 km/20 mi wide; there are usually navigable passes through the barrier into the lagoon. *Atolls* resemble a ring surrounding a lagoon, and do not enclose an island. They are usually formed by the gradual subsidence of an extinct volcano, the coral growing up from where the edge of the island once lay. Coral reefs face danger from increased tourism.

Coriolis effect the effect of the Earth's rotation on the atmosphere and on all objects on the Earth's surface. In the northern hemisphere it causes moving objects and currents to be deflected to the right; in the southern hemisphere it causes deflection to the left. The effect is named after its discoverer, French mathematician Gaspard Coriolis (1792–1843).

The Coriolis effect can be easily observed by watching water go down a plughole – it does not flow directly downwards but spins to the right (clockwise) or to the left (anticlockwise), depending on whether the observer is in the northern or southern hemisphere.

Costa Rica *by 1983 only 17% of the country's natural forest remained; half of the arable land had been cleared for cattle ranching, which led to landlessness, unemployment (except for 2,000 politically* *powerful families), and soil erosion; the massive environmental destruction also caused incalculable loss to the gene pool. Now one of the leading centres of conservation in Latin America, with more than 10% of the country protected by national parks, and tree replanting proceeding at a rate of 150 sq km/60 sq mi per year.*

country park pleasure ground or park, often located near an urban area, providing facilities for the public enjoyment of the countryside. Country parks were introduced in the UK following the 1968 Countryside Act and are the responsibility of local authorities with assistance from the Countryside Commission. They cater for a range of recreational activities such as walking, boating, and horse-riding. See ⇨national park.

Countryside Commission official conservation body created for England and Wales under the Countryside Act 1968. It replaced the National Parks Commission, and by 1980 had created over 160 country parks.

Countryside Council for Wales Welsh nature conservation body formed 1991 by the fusion of the former ⇨Nature Conservancy Council and the Welsh Countryside Commission. It is government-funded and administers conservation and land-use policies within Wales.

crocodile large aquatic carnivorous reptile of the family Crocodiliae, related to alligators, caymans, and ⇨gavials but distinguished from them by a more pointed snout and a notch in the upper jaw into which the fourth tooth of the lower jaw fits. Crocodiles can grow up to 6 m/ 20 ft, and have long, powerful tails that propel them when swimming.

They have remained virtually unchanged for 200 million years and an individual can live for up to 100 years. About a dozen species of crocodiles, all of them endangered, are found in tropical parts of Africa, Asia, Australia, and Central America. Hunted to near-extinction for their leather, they are now protected by restrictions on the trade in their skins.

crop rotation system of regularly changing the crops grown on a piece of land. The crops are grown in a particular order to utilize and add to the nutrients in the soil and to prevent the build-up of insect and fungal pests. Including a legume crop such as peas or beans in the rotation helps build up nitrate in the soil because the roots contain bacteria capable of fixing nitrogen from the air.

A simple seven-year rotation, for example, might include a three-year ley followed by two years of wheat and then two years of barley, before returning the land to temporary grass once more. In this way, the cereal crops can take advantage of the build-up of soil fertility that occurs during the period under grass. In the 18th century, a four-year rotation was widely adopted with autumn-sown cereal, followed by a root crop, then spring cereal, and ending with a leguminous crop. Since then, more elaborate rotations have been devised with two, three, or four successive cereal crops, and with the root crop replaced by a cash crop such as sugar beet or potatoes, or by a legume crop such as peas or beans.

curie former unit (symbol Ci) of radioactivity, equal to 37×10^9 ⇨becquerels. One gram of radium has a radioactivity of about one curie. It was named after French physicists Marie and Pierre Curie.

cyanobacteria (singular *cyanobacterium*) alternative name for ⇨blue-green algae.

cyclamate derivative of cyclohexysulphamic acid, formerly used as an artificial ⇨sweetener, 30 times sweeter than sugar. First synthesized 1937, its use in foods was banned in the USA and the UK from 1970, when studies showed that massive doses caused cancer in rats.

Czech Republic *considered to be the most polluted country in E Europe. Pollution is worst in N Bohemia, which produced 70% of Czechoslovakia's coal and 45% of its coal-generated electricity. Up to 20 times the permissible level of sulphur dioxide is released over Prague, where 75% of the drinking water fails to meet the country's health standards.*

D

dam structure built to hold back water in order to prevent flooding, to provide water for irrigation and storage, and to provide hydroelectric power. The biggest dams are of the earth-and rock-fill type, also called *embankment dams*. Early dams in Britain, built before about 1800, had a core made from puddled clay (clay which has been mixed with water to make it impermeable). Such dams are generally built on broad valley sites. Deep, narrow gorges dictate a *concrete dam*, where the strength of reinforced concrete can withstand the water pressures involved.

Major dams include: Rogun (Tajikistan), the world's highest at 335 m/1,099 ft; New Cornelia Tailings (USA), the world's biggest in volume, 209 million cu m/7.4 billion cu ft; Owen Falls (Uganda), the world's largest reservoir capacity, 204.8 billion cu m/7.2 trillion cu ft; and Itaipu (Brazil/Paraguay), the world's most powerful, producing 12,700 megawatts of electricity.

Although dams can service huge irrigation schemes and are a reliable and cheap source of power, they cause many environmental problems such as the forcible removal of local communities, waterlogging and salinization of land in the area, and loss of habitat. For example, the world's biggest hydroelectric dam and irrigation project which is currently under construction on the Narmada river, central India, has attracted huge protests as it will displace up to a million people and submerge large areas of forest and farmland. Similarly, the Kansa dam in Zimbabwe flooded habitat used by the rhinoceros, one of the world's most endangered mammals.

There is also controversy as to the effectiveness of large dams as the reservoirs tend to fill with silt from upstream. This leads to a gradual reduction in reservoir depth and hence the volume of water held back by the dam which in turn reduces the power delivered by the hydroelectric turbines. As a result of such problems, interest is now turning away from larger dams to concentrate more on smaller-scale projects without massive reservoirs so that hydroelectric power can be generated efficiently with minimal environmental damage.

daminozide (trade name *Alar*) chemical formerly used by fruit growers to make apples redder and crisper. In 1989 a report published in the USA found the consumption of daminozide to be linked with cancer, and the US ⇨Environmental Protection Agency (EPA) called for an end to its use. The makers have now withdrawn it worldwide.

DDT abbreviation for *dichloro-diphenyl-trichloroethane* $(ClC_6H_5)_2$ $CHCHCl_2$ insecticide discovered 1939 by Swiss chemist Paul Müller. It is useful in the control of insects that spread malaria, but resistant strains develop. DDT is highly toxic and persists in the environment and in living tissue. Its use is now banned in most countries, but it continues to be used on food plants in Latin America.

debt something that is owed by a person, organization, or country, usually money, goods, or services. The *national debt* of a country is the total money owed by the national government to private individuals and banks. *International debt* is the money owed by one country to another.

major debtor nations

country	total international debt 1988 ($ billion)	debt as % of GDP 1988
Brazil	114.6	32.3
Mexico	101.6	55.3
Argentina	58.9	69.5
India	57.5	21.3
Indonesia	52.6	63.6
Egypt	50.0	182.6

The danger of the current scale of international debt (the so-called *debt crisis*) is that the debtor country can only continue to repay its existing debts by means of further loans; for the Western countries, there is the possibility of a confidence crisis causing a collapse of the banking system. During the 1980s the concept of the ⇨debt-for-nature swap was developed as one way of tackling the problem.

debt-for-nature swap agreement under which a proportion of a country's debts are written off in exchange for a commitment by the debtor country to undertake projects for environmental protection. Debt-for-nature swaps were set up by environment groups in the 1980s in an attempt to reduce the debt problem of poor countries, while simultaneously promoting conservation.

To date, most debt-for-nature swaps have concentrated on setting aside areas of land, especially tropical rainforest, for protection and have involved private conservation foundations. The first swap took place 1987, when a US conservation group bought $650,000 of Bolivia's national debt from a bank for $100,000, and persuaded the Bolivian government to set aside a large area of rainforest as a nature reserve in exchange for never having to pay back the money owed. Other countries participating in debt-for-nature swaps are the Philippines, Costa Rica, Ecuador, and Poland. However, the debtor country is expected to ensure that the area of land remains adequately protected, and in practice this does not always happen. Such deals have also produced complaints of neocolonialism.

decibel unit (symbol dB) of measure used originally to compare sound intensities and subsequently electrical or electronic power outputs; now also used to compare voltages. An increase of 10 dB is equivalent to a 10-fold increase in intensity or power, and a 20-fold increase in voltage. A whisper has an intensity of 20 dB; 140 dB (a jet aircraft taking off nearby) is the threshold of pain. See also ⇨noise.

deciduous forest woodland area consisting of broad-leaved trees (such as oak) which shed their leaves in winter to reduce water loss and conserve energy. They are the natural vegetation of northern mainland Europe and the British Isles, but have been chopped down to make way for farming, industry, and settlement. Broad-leaved trees grow slowly, reaching maturity 100–200 years after being planted, thus limiting their economic value. Deciduous forests are comparatively richer habitats than ⇨coniferous forests, for instance in the range of flora and fauna sustained.

decomposer organism that breaks down dead matter. Decomposers play a vital role in the ⇨ecosystem by freeing important chemical substances, such as nitrogen compounds, locked up in dead organisms or excrement. They feed on some of the released organic matter, but leave the rest to filter back into the soil or pass in gas form into the atmosphere. The principal decomposers are bacteria and fungi, but earthworms and many other invertebrates are often included in this group. The ⇨nitrogen cycle relies on the actions of decomposers.

decontamination factor in radiological protection, a measure of the effectiveness of a decontamination process. It is the ratio of the original contamination to the remaining radiation after decontamination: 1,000 and above is excellent; 10 and below is poor.

deforestation the cutting down of forest without planting new trees to replace those lost (reafforestation) or allowing the forest to regenerate itself naturally. In tropical forests, such as those in the Amazon basin in South America, deforestation has been severe over the last few decades because of pressures from farmers and developers. Trees have been cut down to provide firewood and building materials and to make way for mining and urban developments.

Many people are concerned about the rate of deforestation as great damage is being done to the habitats of plants and animals. Deforestation also causes fertile soil to be blown away or washed into rivers, leading to ⇨soil erosion and famine, and is thought to be partially responsible for the flooding of lowland areas – for example, in Bangladesh – because trees help to slow down water movement. It may also increase the carbon dioxide content of the atmosphere and intensify the ⇨greenhouse effect, because there are fewer trees absorbing carbon dioxide from the air for photosynthesis.

The first deforestation occurred more than 2,000 years ago in areas surrounding the Mediterranean as wood was increasingly used for fuel, building materials, and the construction of ships. Throughout the next two millennia most of the woodland in Europe was destroyed as such demands increased and new ones, such as the manufacture of paper, arose. The current wave of deforestation in the tropics dates back only 30 years, but even so has reduced the amount of intact forest ecosystem from 34% of total land in the affected areas to 12%. Deforestation in the tropics is especially serious because such forests do not regenerate easily and they are such a rich source of ⇨biodiversity.

demography study of the size, structure, dispersal, and development of human ⇨populations to establish reliable statistics on such factors as birth and death rates, marriages and divorces, life expectancy, and migration. Demography is used to calculate life tables, which give the life expectancy of members of the population by sex and age.

Demography is significant in the social sciences as the basis for government planning in such areas as education, housing, welfare, transport, and taxation. Demographic changes are also important for industry and many businesses. For example, the fall in the number of people aged 10–20 during the 1980s and the first half of the 1990s has led to many school closures, a shrinkage in the potential market for teenage clothes, and a fall in the number of young people available for recruitment into jobs by employers. Equally, the forecast rise in the number of people aged 75+ over the next 20 years will lead to an expansion of demand for accommodation for the elderly.

denitrification process occurring naturally in soil, where bacteria break down ⇨nitrates to give nitrogen gas, which returns to the atmosphere. Denitrification is a vital part of the ⇨nitrogen cycle.

Denmark *pollution of the North Sea is a concern as is the number of waste sites around the country. However, Denmark has pioneered energy conservation schemes in Europe and has also significantly cut back on air pollution thanks to tighter controls and tax incentives.*

denudation natural loss of soil and rock debris, blown away by wind or washed away by running water, that lays bare the rock below. Over millions of years, denudation causes a general lowering of the landscape.

The landscape is constantly being lifted by plate tectonic activity and levelled by weathering and erosion. For example, the Himalayas are being formed by the collision between two plates, yet they are being constantly denuded by glacial erosion and frost shattering. As a result, overall uplift is not as great as it might be.

depressed area region with substandard economic performance, perhaps as a result of a change in industrial structure, such as a decline in manufacturing industry. An example in the UK is Clydeside, where traditional heavy industries have closed because of reduced demand and exhaustion of raw materials (such as coal and iron ore). Depressed areas may be characterized by high unemployment, low-quality housing, and poor educational standards. Government aid may be needed to reverse such decline.

desalination removal of salt, usually from sea water, to produce fresh water for irrigation or drinking. Distillation has usually been the method adopted, but in the 1970s a cheaper process, using certain polymer materials that filter the molecules of salt from the water by reverse osmosis, was developed. Desalination plants are found mainly along the shores of the Middle East where fresh water is in short supply.

desert arid area without sufficient rainfall and, consequently, vegetation to support human life. The term includes the ice areas of the polar regions (known as cold deserts). Almost 33% of Earth's land surface is desert, and this proportion is increasing. Deserts can generally be reclaimed for agricultural use as soon as water is available, because the mineral content is often high. Once pioneer species are established and water supply consistent, organic matter quickly returns, supporting in turn more plants and invertebrates.

The *tropical desert* belts of latitudes from 5° to 30° are caused by the descent of air that is heated over the warm land and therefore has lost its moisture. Other natural desert types are the *continental deserts*, such as the Gobi, that are too far from the sea to receive any moisture; *rain-shadow deserts*, such as California's Death Valley, that lie in the lee of mountain ranges, where the ascending air drops its rain only on the windward slopes; and *coastal deserts*, such as the Namib, where cold ocean currents cause local dry air masses to descend. Desert surfaces are usually rocky or gravelly, with only a small proportion being covered with sand. Deserts can be created by changes in climate, or by the human-aided process of desertification.

desertification creation of deserts by changes in climate or by human-aided processes so that the soil loses its organic content. This creates finer soil particles which are washed or blown away so that all that is left is a rough sand. Desertification can be reversed if the land can be replanted as the roots will then bind the particles of the soil and its organic content can begin to build up again.

The processes leading to desertification include overgrazing, destruction of forest belts, and exhaustion of the soil by intensive cultivation without restoration of fertility – all of which may be prompted by the pressures of an expanding population or by concentration in land ownership. About 135 million people are directly affected by desertification, mainly in Africa, the Indian subcontinent, and South America.

detergent surface-active cleansing agent. The common detergents are made from fats (hydrocarbons) and sulphuric acid, and their long-chain molecules have a type of structure similar to that of soap molecules: a salt group at one end attached to a long hydrocarbon 'tail'. They have an advantage over soap in that they do not produce scum by forming insoluble salts with the calcium and magnesium ions present in hard water. 'Environmentally friendly' detergents do not contain phosphates which cause ⇨eutrophication and ⇨algal bloom in rivers and lakes.

developed world or *First World* or *the North* the countries that have a money economy and a highly developed industrial sector. They generally also have a high degree of urbanization, a complex communications network, high gross domestic product (over $2,000) per person, low birth and death rates, high energy consumption, and a large proportion of the workforce employed in manufacturing or service industries (secondary to quaternary industrial sectors). The developed world includes the USA, Canada, Europe, Japan, Australia, and New Zealand.

developing world or *Third World* or *the South* countries with a largely subsistence economy where the output per person and the average income are low. These countries typically have low life expectancy, high birth and death rates, poor communications and health facilities, low literacy levels, high national debt, and low energy consumption per person. The developing world includes much of Africa and parts of Asia and South America. Terms like 'developing world' and 'less developed countries' are criticized for implying that a highly industrialized economy is a desirable goal.

development the acquisition by a society of industrial techniques and technology; hence the common classification of the 'developed' nations of the First and Second Worlds and the poorer, 'developing' or 'under-developed' nations of the Third World.

The concept has been broadened of late to include improvements in the 'quality of life' – for example, in health care, life expectancy, education, and housing. The assumption that development in the sense of industrialization is inherently good has been increasingly questioned since the 1960s, especially as the environmental consequences of industrial growth have gained a higher public profile.

devil wind minor form of tornado, usually occurring in fine weather; formed from rising thermals of warm air (as is a cyclone). A fire creates a similar updraught.

A *fire devil* or *firestorm* may occur in oil-refinery fires, or in the fire-bombings of cities, for example Dresden, Germany, in World War II.

dew point temperature at which the air becomes saturated with water vapour. At temperatures below the dew point, the water vapour condenses out of the air as droplets. If the droplets are large they become deposited on the ground as dew; if small they remain in suspension in the air and form mist or fog.

diesel engine ⇨internal-combustion engine that burns a lightweight fuel oil. The diesel engine operates by compressing air until it becomes sufficiently hot to ignite the fuel. It is a piston-in-cylinder engine, like the ⇨petrol engine, but only air (rather than an air-and-fuel mixture) is taken into the cylinder on the first piston stroke (down). The piston moves up and compresses the air until it is at a very high temperature. The fuel oil is then injected into the hot air, where it burns, driving the piston down on its power stroke. For this reason the engine is called a compression-ignition engine.

Diesel engines have sometimes been marketed as 'cleaner' than petrol engines because they do not need lead additives and produce fewer gaseous pollutants. However, they do produce high levels of the tiny black carbon particles called particulates, which are believed to be carcinogenic and may exacerbate or even cause asthma.

diet the range of foods eaten, also a particular selection of food, or the

overall intake and selection of food for a particular person or people.
The basic components of a diet are a group of chemicals: proteins, carbohydrates, fats, vitamins, minerals, and water.

variations in calorie intake

percentage of calorie requirement

	1961–63	1969–71	1972–74
developed countries	125	131	134
developing countries	91	92	90
world average	101	106	107

For most of the world's citizens, diet is determined by financial con-
siderations and availability. However, in developed countries fashion
and concern about health issues have had a great influence on people's
dietary habits. An adequate diet is one that fulfills the body's nutritional
requirements and gives an energy intake proportional to the person's
activity level (the average daily requirement is 2,400 calories for men,
less for women, more for active children). In the Third World and in
famine or poverty areas some 450 million people in the world subsist
on fewer than 1,500 calories per day, whereas in the developed countries the average daily intake is 3,300 calories.

NACNE (UK National Advisory Committee on Nutritional Education) guidelines for a healthy diet

component	amount
fat	should be 35% of total energy
fibre	should be 25–30 g per day
protein	should be 10–12% of total energy
cholesterol	all right if fat guidelines are followed
sugar	maximum 55 g per day
salt	maximum 9 g per day

dioxin any of a family of over 200 organic chemicals, all of which are
heterocyclic ⇨hydrocarbons.

The term is commonly applied, however, to only one member of the
family, 2,3,7,8-tetrachlorodibenzo-*p*-dioxin (2,3,7,8-TCDD), a highly

toxic chemical that occurs, for example, as an impurity in the defoliant ⇨Agent Orange, used in the Vietnam War, and sometimes in the weed-killer 2,4,5-T. It has been associated with a disfiguring skin complaint (chloracne), birth defects, miscarriages, and cancer.

Disasters involving accidental release of large amounts of dioxin into the environment have occurred at Seveso, Italy, and Times Beach, Missouri, USA. Small amounts of dioxins are released by the burning of a wide range of chlorinated materials (treated wood, exhaust fumes from fuels treated with chlorinated additives, and plastics) and as a sideeffect of some techniques of paper-making. The possibility of food becoming contaminated by dioxins in the environment has led the EC to significantly decrease the allowed levels of dioxin emissions from incinerators. Dioxin may be produced as a by-product in the manufacture of the bactericide hexachlorophene.

UK government figures released in 1989 showed dioxin levels 100 times higher than guidelines set for environmental dioxin in breast milk, suggesting dioxin contamination is more widespread than previously thought.

dolphin any of various highly intelligent aquatic mammals of the family Delphinidae, which also includes porpoises. There are about 60 species. The name 'dolphin' is generally applied to species having a beaklike snout and slender body, whereas the name 'porpoise' is reserved for the smaller species with a blunt snout and stocky body.

The river dolphins, of which there are only five species, belong to the family Platanistidae. All river dolphins are threatened by dams and pollution, and some, such as the whitefin dolphin *Lipotes vexillifer* of the Chiang Jiang River, China, are in danger of extinction. Marine dolphins are endangered by fishing nets, speedboats, and pollution. In 1990 the North Sea states agreed to introduce legislation to protect them.

Heaviside's dolphin or *benguela dolphin Cephalorhynchus heavisidii* one of the least-known dolphins, confined to the coastal waters of Namibia. It is thought that about 100 a year are killed in purse seine nets from fishing boats, which could endanger this apparently rare species.

Dounreay experimental nuclear reactor site on the north coast of Scotland, 12 km/7 mi W of Thurso. Development started 1974 and continued until the site was decommissioned 1994.

drinking water water that has been subjected to various treatments, including filtration and sterilization, to make it fit for human consumption; it is not pure water.

The EC Drinking Water Directive sets minimum standards for drinking water by specifying 40 important parameters to be monitored including taste, odour, mineral content, nitrate and pesticide content, and bacterial concentration. In the UK, average consumption of water is 140 litres per person per day, some 30% of which is used in flushing the toilet. See also ⇨water and ⇨water supply.

dust bowl area in the Great Plains region of North America (Texas to Kansas) that suffered extensive wind erosion as the result of drought and poor farming practice in once-fertile soil. Much of the topsoil was blown away in the droughts of the 1930s and the 1980s.

Similar dust bowls are being formed in many areas today, noticeably across Africa, because of overcropping and overgrazing.

Dutch elm disease disease of elm trees *Ulmus*, principally Dutch, English, and American elm, caused by the fungus *Certocystis ulmi*. The fungus is usually spread from tree to tree by the elm-bark beetle, which lays its eggs beneath the bark. The disease has no cure, and control methods involve injecting insecticide into the trees annually to prevent infection, or the destruction of all elms in a broad band around an infected area, to keep the beetles out.

The disease was first described in the Netherlands and by the early 1930s had spread across Britain and continental Europe, as well as North America.

In the 1970s a new epidemic was caused by a much more virulent form of the fungus, probably brought to Britain from Canada.

E

eagle any of several genera of large birds of prey of the family Accipitridae, including the golden eagle *Aquila chrysaetos* of Eurasia and North America, which has a 2 m/6 ft wingspan and is dark brown.

The white-headed bald eagle *Haliaetus leucocephalus* is the symbol of the USA; rendered largely infertile through the ingestion of agricultural chemicals, it is now very rare, except in Alaska. Another endangered species is the Philippine eagle, sometimes called the Philippine monkey-eating eagle (although its main prey is flying lemurs). Loss of large tracts of forest, coupled with hunting by humans, have greatly reduced its numbers.

Earth third planet from the Sun. It is almost spherical, flattened slightly at the poles. About 70% of the surface (including the north and south polar icecaps) is covered with water. The Earth is surrounded by a life-supporting atmosphere and is the only planet on which life is known to exist, and is composed of three concentric layers: the core, the mantle, and the crust.

structure the Earth's interior is thought to be composed of a number of concentric layers: an inner core of solid iron and nickel; an outer core of molten iron and nickel; and a mantle of mostly solid rock, separated by the Mohorovičić discontinuity from the Earth's crust.

Evidence for the layered structure has been gathered by scientists surveying the paths taken by seismic waves (earthquake waves), which travel at different speeds through different materials. The crust and the topmost layer of the mantle (the lithosphere) form about 12 large moving plates, some of which carry the continents. The plates are in constant, slow motion, called tectonic drift.

ecology study of the relationships of organisms to each other and to

ecology: chronology

1735	Swedish naturalist Carl Linnaeus developed his system for classifying and naming plants and animals.
1798	English reverend Thomas Malthus produced the earliest theoretical study of population dynamics.
1859	English naturalist Charles Darwin published his theory on *The Origin of the Species*.
1869	German zoologist Ernst Haeckel first defined the term ecology.
1899	US botanist Henry Cowles published his classic paper on succession in sand dunes on Lake Michigan, USA.
1913	British Ecological Society founded.
1915	Ecological Society of America founded.
1916	US ecologist Frederic Clements coined the phrase 'climax communities' for large areas of rather uniform vegetation which he attributed to climactic factors.
1926	Russian botanist N I Vavilov published *Centres of Origin of Cultivated Plants*, concluding that there are relatively few such centres, many of which are located in mountainous areas.
1934	Russian ecologist G F Gause first stated the principles of competitive exclusion, related to a species' niche.
1935	British ecologist Arthur Tansley first coined the term ecosystem.
1938	The coelacanth, a marine fish believed to have become extinct 65 million years ago, was 'rediscovered' in the Indian Ocean.
1940	Population biologist Charles Elton developed the idea of trophic levels in a community of organisms.
1950	The theory that natural selection may favour either individuals with high reproductive rates and rapid development (*r*-selection) or individuals with low reproductive rates and better competitive ability (*k*-selection) is first discussed.
1967	US biologists MacArthur and Wilson proposed their 'Theory of Island Biogeography' which related population and community size to island size. The theory is still widely used in the design of nature reserves today.
1979	English naturalist James Lovelock proposed his Gaia hypothesis, viewing the planet as a single organism.
1993	UN Convention on Biological Diversity came into force.

the environments in which they live, including all living and nonliving components. The term was coined by the biologist Ernst Haeckel 1866.

Ecology may be concerned with individual organisms (for example, behavioural ecology, feeding strategies), with populations (for example, population dynamics), or with entire communities (for example, competition between species for access to resources in an ecosystem, or predator–prey relationships). Applied ecology is concerned with the management and conservation of habitats and the consequences and control of pollution.

ecosystem in ⇨ecology, an integrated unit consisting of the ⇨community of living organisms and the nonliving, or physical, environment in a particular area such as a tropical rainforest, a coral reef or grassland. The relationships among species in an ecosystem are usually complex and finely balanced, and removal of any one species may be disastrous. The removal of a major predator, for example, can result in the destruction of the ecosystem through overgrazing by herbivores. Ecosystems can be identified at different scales – for example, the global ecosystem consists of all the organisms living on Earth, the Earth itself (both land and sea), and the atmosphere above; a freshwater pond ecosystem consists of the plants and animals living in the pond, the pondwater and all the substances dissolved or suspended in that water, and the rocks, mud, and decaying matter that make up the pond bottom.

Energy and nutrients pass through organisms in an ecosystem in a particular sequence (see ⇨food chain): energy is captured through ⇨photosynthesis, and nutrients are taken up from the soil or water by plants; both are passed to herbivores that eat the plants and then to carnivores that feed on herbivores. These nutrients are returned to the soil through the decomposition of excrement and dead organisms, thus completing a cycle that is crucial to the stability and survival of the ecosystem.

Ecuador *rapid population growth has put pressure on all areas in the country, leading to air pollution in urban centres and deforestation in rural areas. This in turn has also led to degradation of the land, particularly as overspill populations from the cities cultivate new areas, encouraged by the government. About 25,000 species became extinct 1965–90 as a result of environmental destruction.*

---

effluent liquid discharge of waste from an industrial process, usually into rivers or the sea, frequently illegally. Effluent is often toxic but is difficult to control and hard to trace. In some cases, as at ⇨Minamata, it can be lethal but usually the toxic effects of effluent remain unclear, because it quickly dilutes in the aquatic ecosystem. For many years ⇨Greenpeace has campaigned against the low-level radioactive effluent discharged by the Sellafield nuclear facility in Cumbria.

Egypt *the building of the Aswan Dam (opened 1970) on the Nile has caused widespread salinization and an increase in waterborne diseases in villages close to Lake Nasser. A dramatic fall in the annual load of silt deposited downstream has reduced the fertility of cropland and has led to coastal erosion and the consequent loss of sardine shoals. Oil pollution from refineries and commercial shipping also poses a significant threat to Egypt's coastal and maritime ecosystems.*

electricity the most useful and most convenient form of energy, readily convertible into heat and light and used to power machines. Electricity can be generated in one place and distributed anywhere because it readily flows through wires. It is generated at power stations where a suitable energy source is harnessed to drive turbines that spin electricity generators. Current energy sources are coal, oil, water power (hydroelectricity), natural gas, and ⇨nuclear energy. Research is under way to increase the contribution of wind, tidal, and geothermal power. Nuclear fuel has proved a more expensive source of electricity than initially anticipated and worldwide concern over radioactivity may limit its future development.

Electricity is generated at power stations at a voltage of about 25,000 volts, which is not a suitable voltage for long-distance transmission. For minimal power loss, transmission must take place at very high voltage (400,000 volts or more). The generated voltage is therefore increased ('stepped up') by a transformer. The resulting high-voltage electricity is then fed into the main arteries of the grid system, an interconnected network of power stations and distribution centres covering a large area. After transmission to a local substation, the line voltage is reduced by a step-down transformer and distributed to consumers. Conventional coal- and oil-fired power stations only convert about 40% of the fuel into electric power and the rest is wasted as heat (but see ⇨combined heat and power stations).

UK electricity generation was split into four companies 1990 in preparation for nationalization. The nuclear power stations remain in the hands of the state through Nuclear Electric (accounting for 20% of electricity generated); National Power (50%) and PowerGen (30%) generate electricity from fossil-fuel and renewable sources. Transmission lines and substations are owned by the National Grid.

electromagnetic pollution The electric and magnetic fields set up by high tension power cables, local electric substations and domestic items such as electric blankets. These electromagnetic fields have been linked to increased levels of cancer, especially leukaemia, and to headaches, nausea, dizziness, and depression.

A considerable amount of research into electromagnetic pollution is now being carried out in the USA and a recently the New York Court of Appeal ruled that landowners living near transmission cables could be awarded damages if it were shown that 'cancer phobia' had lowered the value of their property. In the UK however the issue has so far failed to receive official recognition.

electron stable, negatively charged elementary particle; it is a constituent of all atoms, and a member of the class of particles known as leptons. The electrons in each atom surround the nucleus in groupings called shells; in a neutral atom the number of electrons is equal to the number of protons in the nucleus. This electron structure is responsible for the chemical properties of the atom.

electrostatic precipitator device that removes dust or other particles from air and other gases by electrostatic means. An electric discharge is passed through the gas, giving the impurities a negative electric charge. Positively charged plates are then used to attract the charged particles and remove them from the gas flow. Such devices are attached to the chimneys of coal-burning power stations to remove ash particles.

element substance that cannot be split chemically into simpler substances. The atoms of a particular element all have the same number of protons in their nuclei (their atomic number). Elements are classified in the periodic table of the elements. Of the 109 known elements, 95 are known to occur in nature (those with atomic numbers 1–95). Those from 96 to 109 do not occur in nature and are synthesized only, produced in particle accelerators. Eighty-one of the elements are stable; all the others, which include atomic numbers 43, 61, and from 84 up, are radioactive.

elephant mammal belonging to either of two surviving species of the order Proboscidea: the Asian elephant *Elephas maximus* and the African elephant *Loxodonta africana*. Elephants can grow to 4 m/13 ft and weigh up to 8 tonnes; they have a thick, grey, wrinkled skin, a large head, a long trunk used to obtain food and water, and upper incisors or tusks, which grow to a considerable length. The African elephant has very large ears and a flattened forehead, and the Asian species has smaller ears and a convex forehead. In India, Myanmar (Burma), and Thailand, Asiatic elephants are widely used for transport and logging.

Elephants are slaughtered needlessly for the ⇨ivory of their tusks, and this, coupled with the fact that they reproduce slowly and do not breed readily in captivity, is leading to their extinction. In Africa, over-hunting caused numbers to collapse during the 1980s and the elephant population of E Africa is threatened with extinction. There were 1.3 million African elephants in 1981; fewer than 700,000 in 1988, and about 600,000 in 1990. They were placed on the ⇨CITES list of most endangered species in 1989. A world ban on trade in ivory was imposed 1990, resulting in an apparent drop in poaching, but it is not yet certain what long-term effects the ban will have.

El Niño (Spanish 'the child') warm ocean surge of the Peru Current, so called because it tends to occur at Christmas, recurring every 5–8 years or so in the E Pacific off South America. It involves a change in the direction of ocean currents, which prevents the upwelling of cold, nutrient-rich waters along the coast of Ecuador and Peru, killing fishes and plants. It is an important factor in global weather.

El Niño is believed to be caused by the failure of trade winds and, consequently, of the ocean currents normally driven by these winds. Warm surface waters then flow in from the east. The phenomenon can disrupt the climate of the area disastrously, and has played a part in causing famine in Indonesia, drought and bush fires in the Galàpagos

Islands, rainstorms in California and South America, and the destruction of Peru's anchovy harvest and other marine wildlife 1982–83; algal blooms in Australia's drought-stricken rivers and an unprecedented number of typhoons in Japan 1991; and snowstorms in the USA and bushfires in Australia during which thousands of homes were destroyed and dozens of people killed 1993.

endangered species plant or animal species whose numbers are so few that it is at risk of becoming extinct. Officially designated endangered species are listed by the International Union for the Conservation of Nature (⇨IUCN).

The Javan rhinoceros is the rarest large mammal on Earth – there are only about 60 alive today and, unless active steps are taken to promote this species' survival, it will probably be extinct within a few decades (see also ⇨rhinoceros).

energy capacity for doing work. Potential energy (PE) is energy deriving from position; thus a stretched spring has elastic PE, and an object raised to a height above the Earth's surface, or the water in an elevated reservoir, has gravitational PE. A lump of coal and a tank of petrol, together with the oxygen needed for their combustion, have chemical energy. Other sorts of energy include electrical and nuclear energy, and light and sound. Moving bodies possess kinetic energy (KE). Energy can be converted from one form to another, but the total quantity stays the same (in accordance with the conservation of energy principle). For example, as an apple falls, it loses gravitational PE but gains KE.

Although energy is never lost, after a number of conversions it tends to finish up as the kinetic energy of random motion of molecules (of the air, for example) at relatively low temperatures. This is 'degraded' energy in that it is difficult to convert it back to other forms.

So-called energy resources are stores of convertible energy. Non-renewable resources include the fossil fuels (coal, oil, and gas) and nuclear-fission 'fuels' – for example, uranium-235. Renewable resources, such as wind, tidal, and geothermal power, have so far been less exploited. Hydroelectric projects are well established, and wind turbines and tidal systems are being developed.

energy, alternative energy from sources that are renewable and ecologically safe, as opposed to sources that are nonrenewable with toxic by-products, such as coal, oil, or gas (fossil fuels), and uranium (for nuclear power). At present the most commercially important and exploited alternative energy source is flowing water, harnessed as ⇨hydroelectric power. Other sources include the oceans' tides and waves (see ⇨tidal power station and ⇨wave power), wind (harnessed by windmills and ⇨wind turbines), the Sun (⇨solar energy), and the heat trapped in the Earth's crust (⇨geothermal energy).

The Centre for Alternative Technology, near Machynlleth in mid-Wales, was established 1975 to research and demonstrate methods of harnessing wind, water, and solar energy.

energy conservation methods of reducing energy use through insulation, increasing energy efficiency, and changes in patterns of use. Profligate energy use by industrialized countries contributes greatly to air pollution and the ⇨greenhouse effect when it draws on nonrenewable energy sources. Increased energy conservation can significantly cut both air pollution and industrial costs. The USA, especially California, has pioneered energy conservation measures in both industry and the community.

It has been calculated that increasing energy efficiency alone could reduce carbon dioxide emissions in several high-income countries by 1–2% a year. The average annual decrease in energy consumption in relation to gross national product 1973–87 was 1.2% in France, 2% in the UK, 2.1% in the USA, and 2.8% in Japan.

By applying existing conservation methods, UK electricity use could be reduced by 4 gigawatts by the year 2000 – the equivalent of four Sizewell nuclear power stations – according to a study by the Open University. This would also be cheaper than building new generating plants.

English Nature agency created 1991 from the division of the ⇨Nature Conservancy Council into English, Scottish, and Welsh sections.

E number code number for additives that have been approved for use by the European Commission (EC). The E written before the number stands for European. E numbers do not have to be displayed on lists of ingredients, and the manufacturer may choose to list ⇨additives by their name instead. E numbers cover all categories of additives apart from flavourings. Additives, other than flavourings, that are not approved by the EC, but are still used in Britain, are represented by a code number without an E.

environment the conditions affecting a particular organism, including physical surroundings, climate, and influences of other living organisms. See also ⇨habitat.

In common usage, 'the environment' often means the total global environment, without reference to any particular organism.

environmental audit see ⇨green audit.

environmentalism theory emphasizing the primary influence of the environment on the development of groups or individuals. It stresses the importance of the physical, biological, psychological, or cultural environment as a factor influencing the structure or behaviour of animals, including humans.

In politics this has given rise in many countries to ⇨Green parties, which aim to 'preserve the planet and its people'. A 1989 UK Department of the Environment survey showed public concern about the environment was second only to worries about the National Health Service.

Environmentally Sensitive Area (ESA) scheme introduced by the UK Ministry of Agriculture 1987, as a result of EC legislation, to protect some of the most beautiful areas of the British countryside from the loss and damage caused by agricultural change. The first areas to be designated ESAs are in the Pennine Dales, the North Peak District, the Norfolk Broads, the Breckland, the Suffolk River Valleys, the Test Valley, the South Downs, the Somerset Levels and Moors, West Penwith, Cornwall, the Shropshire Borders, the Cambrian Mountains, and the Lleyn Peninsula.

The total area designated as ESA's was estimated 1993 at 785,600 hectares. The scheme is voluntary, with farmers being encouraged to adapt their practices so as to enhance or maintain the natural features of the landscape and conserve wildlife habitat. A farmer who joins the scheme agrees to manage the land in this way for at least five years. In return for this agreement, the Ministry of Agriculture pays the farmer a sum that reflects the financial losses incurred as a result of reconciling conservation with commercial farming.

Environmental Protection Agency (EPA) US agency set up 1970 to control water and air quality, industrial and commercial wastes, pesticides, noise, and radiation. In its own words, it aims to protect 'the country from being degraded, and its health threatened, by a multitude of human activities initiated without regard to long-ranging effects upon the life-supporting properties, the economic uses, and the recreational value of air, land, and water'.

erosion wearing away of the Earth's surface, caused by the breakdown and transportation of particles of rock or soil (by contrast, ⇨weathering does not involve transportation). Agents of erosion include the sea, rivers, glaciers, and wind. Water, consisting of sea waves and currents, rivers, and rain; ice, in the form of glaciers; and wind, hurling sand fragments against exposed rocks and moving dunes along, are the most potent forces of erosion. People also contribute to erosion by bad farming practices and the cutting down of forests, which can lead to the formation of ⇨dust bowls. There are several processes of erosion including hydraulic action, corrasion, attrition, and solution.

Ethiopia more than 90% of the forests of the Ethiopian highlands have been destroyed since 1900. The effects of severe drought have been exacerbated by ongoing civil wars over the last decade, leading to widespread famine.

eugenics study of ways in which the physical and mental quality of a people can be controlled and improved by selective breeding, and the belief that this should be done. The idea was abused by the Nazi Party in Germany during the 1930s to justify the attempted extermination of entire groups of people. Eugenics can try to control the spread of inherited genetic abnormalities by counselling prospective parents.

In 1986 Singapore adopted an openly eugenic policy by guaranteeing pay increases to female graduates when they give birth to a child, while offering grants towards house purchases for nongraduate women on condition that they are sterilized after the first or second child.

European Atomic Energy Commission (Euratom) organization

established by the second Treaty of Rome 1957, which seeks the coop-
eration of member states of the European Community in nuclear
research and the rapid and large-scale development of nonmilitary
nuclear energy.

eutrophication excessive enrichment of rivers, lakes, and shallow
sea areas, primarily by nitrate fertilizers washed from the soil by rain,
by phosphates from fertilizers, and from nutrients in municipal sewage,
and by sewage itself. These encourage the growth of algae and bacteria
which use up the oxygen in the water, thereby making it uninhabitable
for fish and other animal life.

Waterways close to urban developments or areas of intensive farming are particularly at risk from eutrophication, for example, the Great
Lake in Canada and the Norfolk Broads in the UK.

extinction the complete disappearance of a species. In the past,
extinctions are believed to have occurred because species were unable
to adapt quickly enough to a naturally changing environment. Today,
most extinctions are due to human activity. Some species, such as the
dodo of Mauritius, the moas of New Zealand, and the passenger pigeon
of North America, were exterminated by hunting. Others became
extinct when their habitat was destroyed. See also ⇨endangered species.

Mass extinctions are episodes during which whole groups of species
have become extinct, the best known being that of the dinosaurs, other
large reptiles, and various marine invertebrates about 65 million years
ago. Another mass extinction occurred about 10,000 years ago when
many giant species of mammal died out. This is known as the 'Pleis-
tocene overkill' because their disappearance was probably hastened by
the hunting activities of prehistoric humans. The greatest mass extinction occurred about 250 million years ago marking the Permian–
Triassic boundary, when up to 96% of all living species became extinct.
It was proposed 1982 that mass extinctions occur periodically, at
approximately 26-million-year intervals.

The current mass extinction is largely due to human destruction of
habitats, as in the tropical forests and coral reefs; it is far more serious
and damaging than mass extinctions of the past because of the speed at

THE RACE TO PREVENT MASS EXTINCTION

About a million different living species have been identified so far. Recent studies in tropical forest – where biodiversity is greatest – suggest the true figure is nearer 30 million. Most are animals, and most of those are insects. Because the tropical forests are threatened, at least half the animal species could become extinct within the next century. There have been at least five 'mass extinctions' in our planet's history; the last removed the dinosaurs 65 million years ago. The present wave of extinction is of similar scale, but hundreds of times faster.

There is conflict within the conservation movement. Some believe that *habitats* should be conserved; others prefer to concentrate upon individual *species*.

Habitat protection

Habitat protection has obvious advantages. Many species benefit if land is preserved. Animals need somewhere to live; unless the habitat is preserved it may not be worth saving the individual animal. Habitat protection *seems* cheap; for example, tropical forest can often be purchased for only a few dollars per hectare. Only by habitat protection can we save more than a handful of the world's animals.

But there are difficulties. Even when a protected area is designated a 'national park', its animals may not be safe. All five remaining species of rhinoceros are heavily protected in the wild, but are threatened by poaching. Early in 1991 Zimbabwe had 1,500 black rhino – the world's largest population. Patrols of game wardens shoot poachers on sight. Yet by late 1992, 1,000 of the 1,500 had been poached. In many national parks worldwide, the habitat is threatened by the local farmers' need to graze their cattle.

Computer models and field studies show that wild populations need several hundred individuals to be viable. Smaller populations will eventually become extinct in the wild, because of accidents to key breeding individuals, or epidemics. The big predators need vast areas. One tiger may command hundreds of square kilometres; a viable population needs an area as big as Wales or Holland. Only one of the world's five remaining subspecies of tiger – a population of Bengals in India – occupies an area large enough to be viable. All the rest (Indo-Chinese, Sumatran, Chinese and Siberian) seem bound to die out.

Mosaic

Ecologists now emphasize the concept of *mosaic*. All animals need different things from their habitat; a failure of any one is disastrous. For example, giant pandas feed mainly on bamboo, but give birth in old hollow trees—of which there is a shortage. Nature reserves must either contain all essentials for an animal's life, or else allow access to such areas elsewhere. For many animals in a reserve, these conditions are not fulfilled. Hence year by year, after reserves are created, species become extinct: a process called *species relaxation*. The remaining fauna and flora may be a poor shadow of the original.

Interest is increasing in *captive breeding*, carried out mainly by the world's 800 zoos. Each captive species should include several hundred individuals. Zoos maintain such numbers through *cooperative breeding*, organized regionally and coordinated by the Captive Breeding Specialist Group or the World Conservation Union, based in Minneapolis, Minnesota. Each programme is underpinned by a studbook, showing which individuals are related to which.

Genetic diversity

Conservation breeders aim to maintain *maximum genetic diversity* by encouraging every individual to breed, including those reluctant to breed in captivity; by equalizing family size, so one generation's genes are all represented in the next; and by swapping individuals between zoos, to prevent inbreeding.

Cooperative breeding programmes are rapidly diversifying; by the year 2000 there should be several hundred. They can only make a small impression on the 15 million endangered species, but they can contribute greatly to particular groups of animals, especially the land vertebrates—mammals, birds, reptiles and amphibians. These include most of the world's largest animals, with the greatest impact on their habitats. There are 24,000 species of land vertebrate, of which 2,000 probably require captive breeding to survive.

Captive breeding is not intended to establish 'museum' populations, but to provide a temporary 'lifeboat'. Arabian oryx, California condor, black-footed ferret, red wolf, and Mauritius kestrel are among the creatures saved from extinction and returned to the wild.

which it occurs. Artificial climatic changes and pollution also make it less likely that the biosphere can recover and evolve new species to suit a changed environment. The rate of extinction is difficult to estimate, since most losses occur in the rich environment of the tropical rainforest, where the total number of existent species is not known. Conservative estimates put the rate of loss due to deforestation alone at 4–6,000 species a year. Overall, the rate could be as high as one species an hour, with the loss of one species putting those dependent on it at risk. Australia has the worst record for extinction: 18 mammals have disappeared since Europeans settled there, and 40 more are threatened.

The last mouse-eared bat (*Myotis myotis*) in the UK died 1990. This is the first mammal to have become extinct in the UK for 250 years, since the last wolf was exterminated.

Exxon Corporation the USA's largest oil concern, founded 1888 as the Standard Oil Company (New Jersey), selling petrol under the brand name Esso from 1926 and under the name Exxon in the USA from 1972. The company was responsible for ⇨oil spills in Alaska 1989 and New York harbour 1990.

Under a US government settlement 1991, Exxon was ordered to pay a fine of $100 million plus $900 million damages to repair environmental harm done to the Alaskan shoreline as a result of an oil spill from the *Exxon Valdez*. However, the settlement collapsed May 1991 and legal action against Exxon is likely to continue for several years.

F

factory farming intensive rearing of animals for food, usually on high-protein foodstuffs in confined quarters. Chickens were the first animals to be farmed in this way during the 1940s, for egg production. Antibiotics and growth hormones are commonly used to increase yield, although some countries now restrict this practice because they can persist in the flesh of the animals after slaughter. The emphasis is on productive yield rather than animal welfare so that conditions for the animals are often very poor. For this reason, many people object to factory farming on moral as well as health grounds.

The practice was strongly encouraged by successive governments in Europe during and after World War II as a quick and easy method of increasing food supplies, but by the 1970s problems started to become apparent. Although it is cheap in financial terms, the costs of the process in terms of animal welfare and environmental and health problems (as a result of the increasing use of agricultural chemicals) can be high. For example, in 1993 36% of all reported incidents of water pollution in the UK were from agricultural activities (95% were animal-rearing farms).

poultry Chickens used for egg production (battery hens) and for meat (broiler chickens) have now become separate strains. Egg-laying hens are housed in 'batteries' of cages arranged in long rows. In the course of a year, battery hens average 261 eggs each, whereas for free-range chickens the figure is 199. EC requirements to be enforced 1995 stipulate the minimum floor space per bird in battery units as 450 sq cm/70 sq in; in the UK up to 5 birds are typically housed in cages with floor areas of 2,400 sq cm/372 sq in. Battery hens are prone to a range of afflictions such as malignant tumours, broken/brittle bones, deformed

factory farming The battery-farmed hen lives a life restricted to a small cage. Food and drink is available from the trough attached to the cage. Excreta drops through holes in the floor and can be removed on trays. Eggs roll forward along the sloping floor to allow easy collection.

feet, and cannibalism, although debeaking is commonly employed to combat this. Over 600 million broiler chickens are slaughtered annually in the UK for food, and some 30 million 'spent' battery hens are used every year in baby food and school meals. Turkeys and ducks are also farmed in broiler units.

pigs Intensively farmed pigs are reared in narrow crates and housed in dark sheds, where they will spend their entire lives prior to slaughter. Sows give birth in 'farrowing' crates and are kept there for 1–4 weeks until the piglets are weaned, when they are transferred to dry sow stalls and tethered to the floor.

dairy cattle In order to maximize milk yield, dairy cattle are pregnant for 9 out of every 12 months and are milked for 10 months. The calves produced are taken from the cow at 1–3 days old to be used in veal production, fattened for beef, or themselves used as dairy cattle. Dairy cattle produce five times more milk than a calf could drink as a result of selective breeding and the use of boosting drugs. They are slaughtered at 5–6 years old, compared to the natural lifespan of 20 years.

fish Since salmon farming became popular in the 1970s, some 50 million fish have been produced every year on fish farms in the UK. The waste produced by fish farms can cause ⇨algal bloom and the use of pesticides and other chemicals can contaminate waterways. Other wildlife in the area are also killed in order to protect fish stocks: in 1987 over 2,000 cormorants, 1,000 seals, and 200 herons were killed on Scottish salmon farms, according to government figures; animal welfare groups put the figures even higher.

fallout harmful radioactive material released into the atmosphere in the debris of a nuclear explosion and descending to the surface of the Earth. Such material can enter the food chain, cause ⇨radiation sickness, and last for hundreds of thousands of years (see ⇨half-life).

Fallout is responsible for 0.4% of the average background radiation exposure in the UK environment, of which approximately half is attributable to the ⇨Chernobyl disaster.

family planning spacing or preventing the birth of children. Access to family-planning services is a significant factor in women's health as well as in limiting population growth. If all those women who wished to avoid further childbirth were able to do so, the number of births would be reduced by 27% in Africa, 33% in Asia, and 35% in Latin America; and the number of women who die during pregnancy or childbirth would be reduced by about 50%.

The average number of pregnancies per woman is two in the industrialized countries, compared to six or seven pregnancies per woman in the developing world. According to a World Bank estimate, doubling the annual $2 billion spent on family planning would avert the deaths of 5.6 million infants and 250,000 mothers each year.

fast breeder or *breeder reactor* alternative names for ⇨fast reactor, a type of nuclear reactor.

fast reactor or *fast breeder reactor* ⇨nuclear reactor that makes use of fast neutrons to bring about ⇨fission. Unlike other reactors used by the nuclear-power industry, it has little or no ⇨moderator, to slow down neutrons. The reactor core is surrounded by a 'blanket' of uranium carbide and cooled by liquid sodium. During operation, some of this uranium is converted into plutonium, which can be extracted and later used as fuel.

Fast breeder reactors can extract about 60 times the amount of energy from uranium that thermal reactors do. In the 1950s, when uranium stocks were thought to be dwindling, the fast breeder was considered to be the reactor of the future. Now, however, as new uranium reserves have been found and because of various technical difficulties in construction, development of the fast breeder has slowed.

fecundity the rate at which an organism reproduces, as distinct from its ability to reproduce (⇨fertility). In vertebrates, it is usually measured as the number of offspring produced by a female each year.

feedback general principle whereby the results produced in an ongoing reaction become factors in modifying or changing the reaction; it is the principle used in self-regulating control systems, from a simple thermostat and steam-engine governor to automatic computer-controlled machine tools. A fully computerized control system, in which there is no operator intervention, is called a *closed-loop feedback* system. A system that also responds to control signals from an operator is called an *open-loop feedback* system.

In self-regulating systems, information about what *is* happening in a system (such as level of temperature, engine speed, or size of work-piece) is fed back to a controlling device, which compares it with what *should* be happening. If the two are different, the device takes suitable action (such as switching on a heater, allowing more steam to the engine, or resetting the tools). The idea that the Earth is a self-regulating system, with feedback operating to keep nature in balance, is a central feature of the ⇨Gaia hypothesis.

Fermilab (shortened form of *Fermi National Accelerator Laboratory*) US centre for particle physics at Batavia, Illinois, near Chicago. It is named after Italian–US physicist Enrico Fermi. Fermilab was opened in 1972, and is the home of the Tevatron, the world's most powerful particle accelerator. It is capable of boosting protons and antiprotons to speeds near that of light (to energies of 20 TeV).

fertility an organism's ability to reproduce, as distinct from the rate at which it reproduces (⇨fecundity). Individuals become infertile (unable to reproduce) when they cannot generate gametes (eggs or sperm) or when their gametes cannot yield a viable embryo after fertilization.

fertilizer substance containing some or all of a range of about 20 chemical elements necessary for healthy plant growth, used to compensate for the deficiencies of poor or depleted soil. Fertilizers may be *organic*, for example farmyard manure, composts, bonemeal, blood, and fishmeal; or *inorganic*, in the form of compounds, mainly of nitrogen, phosphate, and potash. Inorganic fertilizer use has increased by between 5 and 10 times during the last 50 years in a drive to increase yields.

Because externally applied fertilizers tend to be in excess of plant requirements and drain away to affect lakes and rivers (see ⇨eutrophication), attention has turned to the modification of crop plants themselves. Plants of the legume family, including beans, clover, and lupin, live in symbiosis with bacteria located in root nodules, which fix nitrogen from the atmosphere. Research is now directed to producing a similar relationship between such bacteria and crops such as wheat.

field study study of ecology, geography, geology, history, archaeology, and allied subjects, in the natural environment as opposed to the laboratory.

The Council for the Promotion of Field Studies was established in Britain 1943, in order to promote a wider knowledge and understanding of the natural environment among the public; Flatford Mill, Suffolk, was the first of its research centres to be opened.

firewood the principal fuel for some 2 billion people, mainly in the Third World. In principle a renewable energy source, firewood is being cut far faster than the trees can regenerate in many areas of Africa and Asia, leading to ⇨deforestation.

In Mali, for example, wood provides 97% of total energy consumption, and deforestation is running at an estimated 9,000 hectares a year. The heat efficiency of firewood can be increased by use of well-designed stoves, but for many people they are either unaffordable or unavailable. With wood for fuel becoming scarcer the UN Food and Agricultural Organization has estimated that by the year 2000 3 billion people worldwide will face chronic problems in cooking.

fishing the harvesting of fish and shellfish from the sea or from fresh water – for example, cod from the North Sea, and carp from the lakes of China and India. Fish are an excellent source of protein for humans, and fish products such as oils and bones are used in industry to produce livestock feed, fertilizers, glues, and drugs. The greatest proportion of the world's catch comes from the oceans.

Between 1950 and 1970, the world fish catch increased by an average of 7% each year. Refrigerated factory ships allowed filleting and processing to be done at sea, and Japan evolved new techniques for locating shoals (by sonar and radar) and catching them (for example, with electrical charges and chemical baits). By the 1970s, ⇨overfishing had led to serious depletion of stocks, and heated confrontations between countries using the same fishing grounds. A partial solution was the extension of fishing limits to 320 km/200 mi. The North Sea countries have experimented with the artificial breeding of fish eggs and release of small fry into the sea. In 1988, overfishing of the NE Atlantic led to hundreds of thousands of starving seals on the northern coast of Norway. A United Nations resolution was passed 1989 to end drift-net fishing (an indiscriminate method) by June 1992.

fission the splitting of a heavy atomic nucleus into two or more major fragments. It is accompanied by the emission of two or three neutrons and the release of large amounts of ⇨nuclear energy.

Fission occurs spontaneously in nuclei of uranium-235, the main fuel used in nuclear reactors. However, the process can also be induced by bombarding nuclei with neutrons because a nucleus that has absorbed a neutron becomes unstable and soon splits. The neutrons released spontaneously by the fission of uranium nuclei may therefore be used in turn to induce further fissions, setting up a ⇨chain reaction that must be controlled if it is not to result in a nuclear explosion.

flue-gas desulphurization process of removing harmful sulphur pollution from gases emerging from a boiler. Sulphur compounds such as sulphur dioxide are commonly produced by burning ⇨fossil fuels, especially coal in power stations, and are the main cause of ⇨acid rain. The process is environmentally beneficial but expensive, adding about 10% to the cost of electricity generation.

fluoridation addition of small amounts of fluoride salts to drinking water by certain water authorities to help prevent tooth decay. Experiments in Britain, the USA, and elsewhere have indicated that a concentration of fluoride of 1 part per million in tap water retards the decay of teeth in children by more than 50%.

The recommended policy in Britain is to add sodium fluoride to the water to bring it up to the required amount, but implementation is up to each local authority.

fluoride negative ion (Fl^-) formed when hydrogen fluoride dissolves in water; compound formed between fluorine and another element in which the fluorine is the more electronegative element.

In parts of India, the natural level of fluoride in water is 10 parts per million. This causes fluorosis, or chronic fluoride poisoning, mottling teeth and deforming bones.

fluorocarbon compound formed by replacing the hydrogen atoms of a hydrocarbon with fluorine. Fluorocarbons are used as inert coatings, refrigerants, synthetic resins, and as propellants in ⇨aerosols and fire extinguishers.

There is concern that the release of fluorocarbons – particularly those containing chlorine (⇨chlorofluorocarbons, CFCs) – depletes the ⇨ozone layer, allowing more ultraviolet light from the Sun to penetrate the Earth's atmosphere, increasing the incidence of skin cancer in humans.

food chain a sequence showing the feeding relationships between organisms in a particular ⇨ecosystem. Each organism depends on the next lowest member of the chain for its food.

Energy in the form of food is shown to be transferred from ⇨autotrophs, or producers, which are principally plants and photosynthetic microorganisms to a series of ⇨heterotrophs, or consumers. The heterotrophs comprise the herbivores, which feed on the producers; carnivores, which feed on the herbivores; and ⇨decomposers, which break down the dead bodies and waste products of all four groups (including their own), ready for recycling. In reality, however, organisms have varied diets, relying on different kinds of foods, so that the food chain is an over-simplification. The more complex *food web*

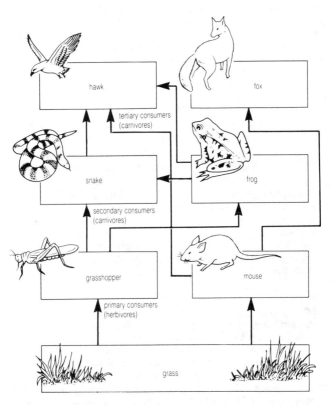

food chain *The complex interrelationships between animals and plants in a food chain. Food chains are normally only three or four links long. This is because most of the energy at each link is lost in respiration, and so cannot be passed up the food chain.*

shows a greater variety of relationships, but again emphasizes that energy passes from plants to herbivores to carnivores. Environmental groups have used the concept of the food chain to show how poisons and other forms of pollution can pass from one animal to another, eventually resulting in the death of rare animals such as the golden eagle *Aquila chrysaetos*.

food technology the commercial processing of foodstuffs in order to render them more palatable or digestible, or to preserve them from spoilage. See ⇨additive and ⇨E number.

forest area where trees have grown naturally for centuries, instead of being logged at maturity (about 150-200 years). A natural, or old-growth, forest has a multistorey canopy and includes young and very old trees (this gives the canopy its range of heights). There are also fallen trees contributing to the very complex ecosystem, which may support more than 150 species of mammals and many thousands of species of insects.

The Pacific forest of the west coast of North America is one of the few remaining old-growth forests in the temperate zone. It consists mainly of conifers and is threatened by logging – less than 10% of the original forest remains.

forestry the large-scale management of trees for commercial or recreational and conservation purposes. Forestry has often been confined to the planting of a single species, such as a rapid-growing conifer providing softwood for paper pulp and construction timber, for which world demand is greatest. It is an example of a primary industry.

In tropical countries, rapid and unmanaged ⇨deforestation has resulted in the destruction of large areas of rainforest, causing environmental problems locally and possibly contributing to ⇨global warming.

fossil fuel fuel, such as coal, oil, and natural gas, formed from the fossilized remains of plants that lived hundreds of millions of years ago. Fossil fuels are a ⇨nonrenewable resource and will eventually run out. Extraction of coal and oil causes considerable environmental pollution, and burning coal contributes to problems of ⇨acid rain by releasing sulphur dioxide.

All fossil fuels release carbon dioxide when burnt, whatever antipollution measures are in place. Increased levels of carbon dioxide are liable to cause ⇨global warming by artificially enhancing the ⇨greenhouse effect.

France as in most Western nations, air pollution in cities due to the combined effects of vehicle emissions and industry is severe. Many rivers are also polluted by effluent from industry and the overuse of agrochemicals in farming districts. France is relatively moderately effected by acid rain compared to other European countries.

free radical an atom or molecule that has an unpaired electron and is therefore highly reactive. Most free radicals are very short-lived. If free radicals are produced in living organisms they can be very damaging.

The action of ultraviolet radiation from the Sun splits ⇨chlorofluorocarbon (CFC) molecules in the upper atmosphere into free radicals, which then break down the ⇨ozone layer.

freeze-drying method of preserving food. The product to be dried is frozen and then put in a vacuum chamber that forces out the ice as water vapour, a process known as sublimation.

Many of the substances that give products such as coffee their typical flavour are volatile, and would be lost in a normal drying process because they would evaporate along with the water. In the freeze-drying process these volatile compounds do not pass into the ice that is to be sublimed, and are therefore largely retained.

Friends of the Earth (FoE or FOE) environmental pressure group, established in the UK 1971, that aims to protect the environment and to promote rational and sustainable use of the Earth's resources. It campaigns on issues such as acid rain; air, sea, river, and land pollution; recycling; disposal of toxic wastes; nuclear power and renewable energy; the destruction of rainforests; pesticides; and agriculture. FoE has branches in 30 countries.

fuel any source of heat or energy, embracing the entire range of materials that burn (combustibles). A *nuclear fuel* is any material that produces energy by nuclear fission in a ⇨nuclear reactor.

fungicide any chemical ⇨pesticide used to prevent fungus diseases in plants and animals. Inorganic and organic compounds containing sulphur are widely used.

fur the hair of certain animals. Fur is an excellent insulating material and so has been used as clothing, although this is criticized by many groups on humane grounds. The methods of farming or trapping animals are often cruel. Mink, chinchilla, and sable are among the most valuable, the wild furs being finer than the farmed. More than 40 million animals are bred in fur farms worldwide every year; in the UK some 55 mink farms and 6 fox farms produce more than 250,000 skins annually. Fur such as mink is made up of a soft, thick, insulating layer called underfur and a top layer of longer, lustrous guard hairs.

Furs have been worn since prehistoric times and have long been associated with status and luxury (ermine traditionally worn by royalty, for example), except by certain ethnic groups such as the Inuit. The fur trade had its origin in North America, exploited by the Hudson's Bay Company from the late 17th century. The chief centres of the fur trade are New York, London, St Petersburg, and Kastoria in Greece. It is illegal to import furs or skins of endangered species listed by ⇨CITES, for example the leopard. Many synthetic fibres are used as substitutes.

G

Gaia hypothesis theory that the Earth's living and nonliving systems form an inseparable whole that is regulated and kept adapted for life by living organisms themselves. The planet therefore functions as a single organism, or a giant cell. Since life and environment are so closely linked, there is a need for humans to understand and maintain the physical environment and living things around them. The Gaia hypothesis was elaborated by British scientist James Lovelock (1919–　) in the 1970s.

gamma radiation very high-frequency electromagnetic radiation, similar in nature to X-rays but of shorter wavelength, emitted by the nuclei of radioactive substances during decay or by the interactions of high-energy electrons with matter. Cosmic gamma rays have been identified as coming from pulsars, radio galaxies, and quasars, although they cannot penetrate the Earth's atmosphere.

Gamma rays are stopped only by direct collision with an atom and are therefore very penetrating; they can, however, be stopped by about 4 cm/1.5 in of lead or by a very thick concrete shield. They are less ionizing in their effect than alpha and beta particles, but are dangerous nevertheless because they can penetrate deeply into body tissues such as bone marrow. They are not deflected by either magnetic or electric fields.

Gamma radiation is used to kill bacteria and other microorganisms, sterilize medical devices, and change the molecular structure of plastics to modify their properties (for example, to improve their resistance to heat and abrasion).

gas-cooled reactor type of nuclear reactor; see ⇨advanced gas-cooled reactor.

gavial large reptile *Gavialis gangeticus* related to the crocodile. It grows to about 7 m/23 ft long, and has a very long snout with about 100 teeth in its jaws. Gavials live in rivers in N India, where they feed on fish and frogs. They have been extensively hunted for their skins, and are now extremely rare.

Geiger counter any of a number of devices used for detecting nuclear radiation and/or measuring its intensity by counting the number of ionizing particles produced (see ⇨radioactivity). It detects the momentary current that passes between ⇨electrodes in a suitable gas when a nuclear particle or a radiation pulse causes the ionization of that gas. The electrodes are connected to electronic devices that enable the number of particles passing to be measured. The increased frequency of measured particles indicates the intensity of radiation.

gene bank collection of seeds or other forms of genetic material, such as tubers, spores, bacterial or yeast cultures, live animals and plants, frozen sperm and eggs, or frozen embryos. These are stored for possible future use in agriculture, plant and animal breeding, or in medicine, ⇨genetic engineering, or the restocking of wild habitats where species have become extinct. Gene banks will be increasingly used as the rate of extinction increases, depleting the Earth's genetic variety (⇨biodiversity).

genetic engineering deliberate manipulation of genetic material by biochemical techniques. It is often achieved by the introduction of new DNA, usually by means of a virus or plasmid. This can be for pure research or to breed functionally specific plants, animals, or bacteria. These organisms with a foreign gene added are said to be transgenic.

In genetic engineering, the splicing and reconciliation of genes is used to increase knowledge of cell function and reproduction, but it can also achieve practical ends. For example, plants grown for food could be given the ability to fix nitrogen, found in some bacteria, and so reduce the need for expensive fertilizers, or simple bacteria may be modified to produce rare drugs. A foreign gene can be inserted into laboratory cultures of bacteria to generate commercial biological products, such as synthetic insulin, hepatitis-B vaccine, and interferon.

Developments in genetic engineering have led to the production of growth hormone, and a number of other bone-marrow stimulating hormones. New strains of animals have also been produced; a new strain of mouse was patented in the USA 1989 (the application was rejected in the European patent office). A vaccine against a sheep parasite (a larval tapeworm) has been developed by genetic engineering; most existing vaccines protect against bacteria and viruses. There is a risk that when transplanting genes between different types of bacteria (*Escherichia coli*, which lives in the human intestine, is often used) new and harmful strains might be produced. For this reason strict safety precautions are observed, and the altered bacteria are disabled in some way so they are unable to exist outside the laboratory.

genome the total information carried by the genetic code of a particular organism.

genotype the particular set of alleles (variants of genes) possessed by a given organism. The term is usually used in conjunction with phenotype, which is the product of the genotype and all environmental effects.

Georgia *heavy pollution of the Black Sea has led to widespread health problems, with much of the Georgian population affected by digestive diseases. Much of the land is also contaminated by toxic pesticides.*

geothermal energy energy extracted for heating and electricity generation from natural steam, hot water, or hot dry rocks in the Earth's crust. Water is pumped down through an injection well where it passes through joints in the hot rocks. It rises to the surface through a recovery well and may be converted to steam or run through a heat exchanger. Dry steam may be directed through turbines to produce electricity. It is an important source of energy in volcanically active areas such as Iceland and New Zealand.

Germany *E Germany suffers chronic pollution due to heavy industrialization in the Communist era. However, the same inefficiencies that led many of these industrial plants to be more polluting than their Western counterparts also means they are unlikely to survive in the free market, thus improving the environmental picture. Acid rain causing ⇨Waldsterben (tree death) affects more than half the country's forests;*

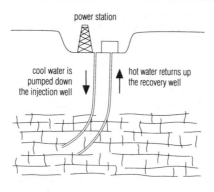

power station

cool water is
pumped down
the injection well

hot water returns up
the recovery well

underground joint system forming underground
resevoir where water is heated

geothermal energy *Geothermal energy arises because the temperature within the Earth's crust increases with depth. On average the temperature rises by 25°C for each kilometre down.*

industrial E Germany has the highest sulphur-dioxide emissions in the world per head of population.

Ghana *deforestation and desertification are widespread. Forested areas shrank from 8.2 million sq km/3.17 million sq mi at the beginning of the 20th century to 1.9 million sq km/730,000 sq mi by 1990. Water pollution is chronic and many of the measures introduced to conserve supplies of water have not tackled contamination and so have served to spread disease. Wildlife conservation measures do exist but are poorly enforced, so that many rare species are under threat, even in reserves.*

global warming projected imminent climate change attributed to the ⇨greenhouse effect. It is estimated that by 2030 the average global temperature will have risen by between 0.7 and 2.0°C. This would raise the average levels of the seas due to the world's oceans expanding as they warm and possible melting of the edges of the polar ice caps, creating a substantial risk of low-lying areas and even entire countries being completely and permanently flooded in many areas. Climate change on this scale would also have a dramatic impact on biodiversity in different areas and crops could be damaged by climate change. In the UK, wetland and peatland areas as well as many coastal communities are reckoned to be most under threat. However, predictions about global warming and its possible climatic effects are tentative and often conflict with each other.

gorilla largest of the apes, *Gorilla gorilla*, found in the dense forests of West Africa and mountains of central Africa. There are three races of the one species: western lowland, eastern lowland, and ⇨mountain gorillas. The male stands about 1.8 m/6 ft, and weighs about 200 kg/450 lb. Females are about half the size. The body is covered with blackish hair, silvered on the back in older males. Gorillas live in family groups; they are vegetarian, highly intelligent, and will attack only in self-defence. They are dwindling in numbers, being shot for food by some local people, or by poachers taking young for zoos, but protective measures are having some effect.

Gran Carajas industrial and mining project in the Brazilian Amazon region, covering an area the size of Britain and France combined. Mining and dam building are destroying huge areas of rainforest and some of the factories are being powered by charcoal from firewood, further adding to the deforestation.

gray SI unit (symbol Gy) of absorbed radiation dose. It replaces the rad (1 Gy equals 100 rad), and is defined as the dose absorbed when one kilogram of matter absorbs one joule of ionizing radiation. Different types of radiation cause different amounts of damage for the same absorbed dose; the SI unit of *dose equivalent* is the ⇨sievert.

Greece acid rain and other airborne pollutants are destroying the classical buildings and ancient monuments of Athens. Air pollution is so severe that the number of vehicles permitted in the city on specified weekdays is restricted and nonessential vehicles have been banned altogether for short periods in an attempt to cut down the build-up of pollutants. Untreated industrial effluent discharged straight into the sea has led to severe pollution in many of the coastal gulfs.

green accounting the inclusion of the economic losses caused by environmental degradation in traditional profit and loss accounting systems. The idea arose in the 1980s when financial factors, and in particular profitability, were the main tool for judging the value of an action. By such crude measures, killing elephants for ivory, destroying tropical rainforest for hardwood, and continuing whaling, all make economic sense. However, if the future value of these resources are included, so that for example tourism, protection of biodiversity, and ecosystem stability are all given a notional economic value, it becomes plain that even in purely economic terms it makes no sense to destroy habitats or hunt animals to extinction.

green audit inspection of a company to assess the total environmental impact of its activities or of a particular product or process. For example, a green audit of a manufactured product looks at the impact of production (including energy use and the extraction of raw materials used in manufacture), use (which may cause pollution and other hazards), and disposal (potential for recycling, and whether waste causes pollution).

Such 'cradle-to-grave' surveys allow a widening of the traditional scope of economics by ascribing costs to variables that are usually ignored such as despoliation of the countryside or air pollution.

green belt area surrounding a large city, officially designated not to be built upon but preserved where possible as open space (for agricultural and recreational use). In the UK the first green belts were introduced in 1947 around conurbations such as London in order to prevent ⇨urban sprawl. New towns were set up to take the overspill population.

green computing the gradual movement by computer companies toward incorporating energy-saving measures in the design of systems and hardware. The increasing use of energy-saving devices, so that a computer partially shuts down during periods of inactivity, but can reactivate at the touch of a key, could play a significant role in ⇨energy conservation. It is estimated that worldwide electricity consumption by computers amounts to 240 billion kilowatt hours per year, equivalent to the entire annual consumption of Brazil. In the USA, carbon dioxide emissions could be reduced by 20 million tonnes per year – equivalent to the carbon dioxide output of 5 million cars – if all computers incorporated the latest 'sleep technology'.

Although it was initially predicted that computers would mean 'paperless offices', in practice the amount of paper consumed continues to rise. Other environmentally-costly features of computers include their rapid obsolescence, health problems concerning monitors and keyboards, and the unfavourable economics of component recycling.

green consumerism catch-phrase especially popular during the 1980s when consumers became increasingly concerned about the environment. Labels such as 'eco-friendly' became a common marketing tool as companies attempted to show that their goods had no negative effect on the environment. The trend was heavily criticized by many environmentalists who argued that it was little more than a conscience-salving exercise, implying that minor changes to lifestyle can save the world.

green tax taxes levied against companies and individuals causing pollution. For example, a company emitting polluting gases or damaging the environment in some other way, would be obliged to pay a correspondingly significant tax; a company which cleans its emissions, reduces its effluent and uses energy-efficient distribution systems, would then be taxed much less.

The idea is often criticized for relying on coercion rather than moral responsibility, but remains high on the political agenda, not least because of the mixed success of antipollution legislation. This practice of using financial rather than primarily legal measures to help the environment has been tried in several US communities: when bin-liners were priced at $1.50 each, the volume of household waste fell dramatically. See also ⇨Clean Air Act.

greenhouse effect phenomenon of the Earth's atmosphere by which the energy of solar radiation, absorbed by the ground and re-emitted as infrared energy, is prevented from escaping by various gases in the air. This results in a rise in the Earth's temperature. Without this effect the Earth would be frozen and uninhabitable. However, the increasing

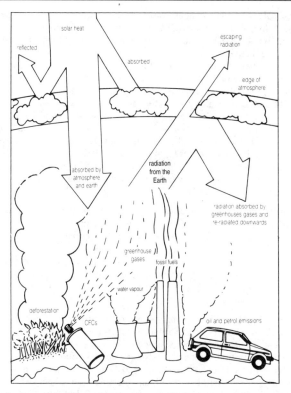

greenhouse effect *The warming effect of the Earth's atmosphere is called the greenhouse effect. Radiation from the Sun enters the atmosphere but is prevented from escaping back into space by gases such as carbon dioxide (produced, for example, by the burning of fossil fuels), nitrogen oxides (from car exhausts), and CFCs (from aerosols and refrigerators). As these gases build-up in the atmosphere, the Earth's average temperature is expected to rise.*

levels of so-called greenhouse gases – such as carbon dioxide, methane, and ⇨chlorofluorocarbons (CFCs) – are enhancing the effect to such an extent that ⇨global warming and dramatic climate change are becoming more likely.

Fossil-fuel consumption and forest fires are the main causes of carbon-dioxide build-up; methane is a byproduct of agriculture (rice, cattle, sheep). Water vapour is another greenhouse gas.

The concentration of carbon dioxide in the atmosphere is estimated to have risen by 25% since the Industrial Revolution, and 10% since 1950; the rate of increase is now 0.5% a year. Chlorofluorocarbon levels are rising by 5% a year, and nitrous oxide levels by 0.4% a year, resulting in a global warming effect of 0.5% since 1900, and a rise of about 0.1°C/3° F a year in the temperature of the world's oceans during the 1980s. Arctic ice was 6–7 m/20–23 ft thick in 1976 and had reduced to 4–5 m/13–17 ft by 1987.

green movement collective term for the individuals and organizations involved in efforts to protect the environment. The movement encompasses political parties such as the ⇨Green Party and organizations like ⇨Friends of the Earth and ⇨Greenpeace.

Despite a rapid growth of public support, and membership of environmental organizations running into many millions worldwide, political green parties have generally failed to win significant levels of the vote in democratic societies.

Green Party political party aiming to 'preserve the planet and its people', based on the premise that incessant economic growth is unsustainable. The leaderless party structure reflects a general commitment to decentralization. Green parties sprang up in W Europe in the 1970s and in E Europe from 1988. Parties in different countries are linked to one another but unaffiliated with any pressure group. The party had a number of parliamentary seats in 1992: Austria 9, Belgium 13, Finland 8, Italy 20, Luxembourg 2, Republic of Ireland 1, Greece 1, and Germany 2; and 29 members in the European Parliament (Belgium 3, France 8, Italy 7, the Netherlands 2, Spain 1, and Germany 8).

The British Green Party was founded 1973 as the Ecology Party (initially solely environmental). In the 1989 European elections, the British

Green Party polled over 2 million votes but received no seats in Parliament, because Britain was the only country in Europe not to have some form of proportional representation. Internal disagreements from 1990 have reduced its effectiveness and popular appeal.

Greenpeace international environmental pressure group, founded 1971, with a policy of nonviolent direct action backed by scientific research. During a protest against French atmospheric nuclear testing in the S Pacific 1985, its ship *Rainbow Warrior* was sunk by French intelligence agents, killing a crew member.

green revolution in agriculture, a popular term for the change in methods of arable farming in Third World countries. The intent is to provide more and better food, albeit with a heavy reliance on chemicals and machinery. It was instigated in the 1940s and 1950s, but abandoned by some countries in the 1980s. Much of the food produced is exported as ⇨cash crops, so that local diet does not always improve.

The green revolution tended to benefit primarily those landowners who could afford the investment necessary for such intensive agriculture. Without a dosage of 70–90 kg/154–198 lb of expensive nitrogen fertilizers per hectare, the high-yield varieties will not grow properly. Hence, rich farmers tend to obtain bigger yields while smallholders are unable to benefit from the new methods.

In terms of production, the green revolution was initially successful in SE Asia; India doubled its wheat yield in 15 years, and the rice yield in the Philippines rose by 75%. However, yields have levelled off in many areas; some countries that cannot afford the dams, fertilizers, and machinery required, have adopted ⇨intermediate technologies.

groundwater water collected underground in porous rock strata and soils; it emerges at the surface as springs and streams. The groundwater's upper level is called the *water table*. Sandy or other kinds of beds that are filled with groundwater are called ⇨aquifers. Recent estimates are that usable ground water amounts to more than 90% of all the fresh water on Earth; however, keeping supplies free from pollutants such as nitrates and pesticides entering the recharge areas is a critical environmental concern.

Most groundwater near the surface moves slowly through the ground while the water table stays in the same place. The depth of the water table reflects the balance between the rate of infiltration, called recharge, and the rate of discharge at springs or rivers or pumped water wells. The force of gravity makes underground water run 'downhill' underground just as it does above the surface. The greater the slope and the permeability, the greater the speed. Velocities vary from 100 cm/ 40 in per day to 0.5 cm/0.2 in.

The water table dropped in the UK during the drought of the early 1990s; however, under cities such as London and Liverpool the water table rose because the closure of industries meant that less water was being removed.

Guatemala *deforestation is becoming a major problem; between 1960 and 1980 nearly 57% of the country's forest was cleared for farming. This is also threatening Guatemala's exceptionally rich biodiversity.*

---

guild a group of species that exploit the same class of environmental resources in a similar way. The group usually numbers about seven, and also applies to human beings, for example, the small fighting units of the Roman army.

H

habitat localized ⇨environment in which an organism lives, and which provides for all (or almost all) of its needs. The diversity of habitats found within the Earth's ecosystem is enormous, and they are changing all the time. Many can be considered inorganic or physical, for example the Arctic ice cap, a cave, or a cliff face. Others are more complex, for instance a woodland or a forest floor. Some habitats are so precise that they are called *microhabitats*, such as the area under a stone where a particular type of insect lives. Most habitats provide a home for many species.

half-life during ⇨radioactive decay, the time in which the strength of a radioactive source decays to half its original value. In theory, the decay process is never complete and there is always some residual radioactivity. For this reason, the half-life of a radioactive isotope is measured, rather than the total decay time. It may vary from millionths of a second to billions of years.

Radioactive substances decay exponentially; thus the time taken for the first 50% of the isotope to decay will be the same as the time taken by the next 25%, and by the 12.5% after that, and so on. For example, carbon-14 takes about 5,730 years for half the material to decay; another 5,730 for half of the remaining half to decay; then 5,730 years for half of that remaining half to decay, and so on. Plutonium-239, one of the most toxic of all radioactive substances, has a half-life of about 24,000 years.

halon organic chemical compound containing one or two carbon atoms, together with bromine and other halogens. The most commonly used are halon 1211 (bromochlorodifluoromethane) and halon 1301 (bromotrifluoromethane). The halons are gases and are widely used in fire extinguishers. As destroyers of the ⇨ozone layer, they are up to ten times more effective than ⇨chlorofluorocarbons (CFCs), to which they are chemically related.

Levels in the atmosphere are rising by about 25% each year, mainly through the testing of fire-fighting equipment.

half-lives of some radioactive isotopes

isotope	half-life
(least stable)	
lithium-5	4.4×10^{-22} sec
polonium-213	4.2×10^{-6} sec
lead-211	36 min
lead-209	3.3 hours
uranium-238	4.551×10^9 years
thorium-232	1.39×10^{10} years
tellurium-128	1.5×10^{24} years
(most stable)	

hazardous substance waste substance, usually generated by industry, which represents a hazard to the environment or to people living or working nearby. Examples include radioactive wastes, acidic resins, arsenic residues, residual hardening salts, lead, mercury, nonferrous sludges, organic solvents, and pesticides. Their economic disposal or recycling is the subject of research.

The UK imported 41,000 tonnes of hazardous waste for disposal 1989, according to official estimates (which exclude chlorinated solvents and nonferrous metals), the largest proportion of which came from Europe. ⇨Greenpeace estimates that in 1991 Britain imported 50,000 tonnes from Europe alone. Most of the annually imported wastes are 'landfilled' (disposed of in dumps) without pretreatment.

heat island large town or city that is warmer than the surrounding countryside. The difference in temperature is most pronounced during the winter, when the heat given off by the city's houses, offices, factories, and vehicles raises the temperature of the air by several degrees.

heavy metal a metallic element of high relative atomic mass, such as platinum, gold, and lead. Many heavy metals are poisonous and tend to *hazard labels* Internationally recognized hazard labels must be attached to lorries carrying dangerous chemicals.

harmful/irritant

toxic

radioactive

explosive

flammable

corrosive

oxidizing/supports fire

accumulate and persist in living systems – for example, high levels of mercury (from industrial waste and toxic dumping) accumulate in shellfish and fish, which are in turn eaten by humans. Treatment of heavy-metal poisoning is difficult because available drugs are not able to distinguish between the heavy metals that are essential to living cells (zinc, copper) and those that are poisonous.

hedge or *hedgerow* row of closely planted shrubs or low trees, generally acting as a land division and windbreak. Hedges also serve as a source of food and as a refuge for wildlife, and provide a ⇨habitat not unlike the understorey of a natural forest.

Between 1945 and 1985, 25% of Britain's hedgerows were destroyed, a length that would stretch seven times around the equator. Since 1989, the UK has suffered a net loss of 26,000 km/16,000 mi of hedgerow. A hedge is a mixture of plants growing together in association – the older the hedge, the greater the number of plant species contained. Because of this ecological richness, the hedge is an important habitat for both vertebrates and invertebrates. Even if hedges are not cut down, mismanagement or the use of pesticides in a neighbouring field can effectively destroy a hedge's value.

herbicide any chemical used to destroy plants or check their growth. See ⇨weedkiller.

hertz SI unit (symbol Hz) of frequency (the number of repetitions of a regular occurrence in one second). Radio waves are often measured in megahertz (MHz), millions of hertz, and the clock rate of a computer is usually measured in megahertz.

heterotroph any living organism that obtains its energy from organic substances produced by other organisms. All animals and fungi are heterotrophs, and they include herbivores, carnivores, and saprotrophs (those that feed on dead animal and plant material).

high-yield variety crop that has been specially bred or selected to produce more than the natural varieties of the same species. During the 1950s and 1960s, new strains of wheat and maize were developed to reduce the food shortages in poor countries (the ⇨green revolution). Later, IR8, a new variety of rice that increased yields by up to six times, was developed in the Philippines. Strains of crops resistant to drought and disease were also developed. High-yield varieties require large amounts of expensive artificial fertilizers and sometimes pesticides for best results.

homeostasis maintenance of a constant internal state in an organism, particularly with regard to pH, salt concentration, temperature, and blood sugar levels. Stable conditions are important for the efficient functioning of the enzyme reactions within the cells, which affect the performance of the entire organism. Within the environmental sciences, the concept can be used to describe the balance of nature, in which the various components of an ecosystem operate to produce a steady state.

hum, environmental disturbing sound of frequency about 40 Hz, heard by individuals sensitive to this range, but inaudible to the rest of the population. It may be caused by industrial noise pollution or have a more exotic origin, such as the jet stream, a fast-flowing high-altitude (about 15,000 m/50,000 ft) mass of air.

Human Genome Project research scheme, begun 1988, to map the complete nucleotide sequence of human DNA. There are approximately 80,000 different genes in the human genome, and one gene may contain more than 2 million nucleotides. The knowledge gained is expected to help prevent or treat many crippling and lethal diseases, but there are potential ethical problems associated with knowledge of an individual's genetic make-up, and fears that it will lead to genetic engineering.

The Human Genome Organization (HUGO) coordinating the project expects to spend $1 billion over the first five years, making this the largest research project ever undertaken in the life sciences. Work is being carried out in more than 20 centres around the world. By the beginning of 1991, some 2,000 genes had been mapped. Concern that, for example, knowledge of an individual's genes may make that person an unacceptable insurance risk has led to planned legislation on genome privacy in the USA, and 3% of HUGO's funds have been set aside for researching and reporting on the ethical implications of the project.

Hungary *an estimated 35%–40% of the population live in areas with officially 'inadmissible' air and water pollution. In Budapest lead levels have reached 30 times the maximum internationally acceptable* *standards. Since the early 1980s, the Hungarian government has had some success in reducing pollution in lakes, an important factor for tourism and leisure.*

hydrocarbon any of a class of chemical compounds containing only hydrogen and carbon (for example, the alkanes and alkenes). Hydrocarbons are obtained industrially principally from petroleum and coal tar. The hydrocarbons contained in car exhaust fumes are believed to be carcinogenic and to contribute to photochemical ⇨smog.

hydroelectric power (HEP) electricity generated by moving water. In a typical HEP scheme, water stored in a reservoir, often created by damming a river, is piped into water ⇨turbines, coupled to electricity generators. In ⇨pumped storage plants, water flowing through the turbines is recycled. A ⇨tidal power station exploits the rise and fall of the tides. About one-fifth of the world's electricity comes from HEP.

HEP plants have prodigious generating capacities. The Grand Coulee plant in Washington State, USA, has a power output of some 10,000 megawatts. The Itaipu power station on the Paraná River (Brazil/Paraguay) has a potential capacity of 12,000 megawatts.

HEP is an example of a ⇨renewable resource. In the UK it provides 1.3% of total electricity production; stations are located in the mountainous areas of Scotland and North Wales where there is fast-flowing water.

hydrogen bomb bomb that works on the principle of ⇨nuclear fusion. Large-scale explosion results from the thermonuclear release of energy when hydrogen nuclei are fused to form helium nuclei. The first hydrogen bomb was exploded at Eniwetok Atoll in the Pacific Ocean by the USA 1952.

hydrogen cyanide HCN poisonous gas formed by the reaction of sodium cyanide with dilute sulphuric acid; it is used for fumigation. It is believed that hydrogen cyanide was present in large quantities in the early atmosphere.

The salts formed from it are cyanides – for example sodium cyanide, used in hardening steel and extracting gold and silver from their ores. If dissolved in water, hydrogen cyanide gives hydrocyanic acid.

hydrogen sulphide H_2S poisonous gas with the smell of rotten eggs. It is found in certain types of crude oil where it is formed by decomposition of sulphur compounds and is sometimes produced by ⇨anaerobic bacteria. It is removed from the oil at the refinery and converted to elemental sulphur.

hydrology study of the location and movement of inland water, both frozen and liquid, above and below ground. It is applied to major civil engineering projects such as irrigation schemes, dams, and hydroelectric power, and in planning water supply.

hydrolysis a form of ⇨chemical weathering caused by the chemical alteration of certain minerals as they react with water. For example, the mineral feldspar in granite reacts with water to form a white clay called china clay.

hydrosphere the water component of the Earth, usually encompassing the oceans, seas, rivers, streams, swamps, lakes, groundwater, and atmospheric water vapour.

Hz in physics, the symbol for ⇨*hertz*.

I

IAEA abbreviation for ⇨International Atomic Energy Agency.

ibis any of various wading birds, about 60 cm/2 ft tall, in the same family, Threskiornidae, as spoonbills. Ibises have long legs and necks, and long, curved beaks. Various species occur in the warmer regions of the world. The Japanese ibis is in danger of extinction because of loss of its habitat; fewer than 25 birds remain.

incineration disposal of waste by burning, a suggested alternative method of ⇨waste disposal to the use of ⇨landfill sites. However incineration is itself far from ideal. It is distrusted by the public because incinerators are considered smelly and poisonous and environmental groups often oppose incineration, partly because of fears of pollution, partly because they consider it better to recycle or reuse the goods. Proponents of incineration claim that the new generation of incinerators, burning at 1000C/1,800°F are efficient and nonpolluting, and produce enough heat to make electricity generation feasible.

India *the controversial Narmada Valley Project is the world's largest combined hydroelectric irrigation scheme. In addition to displacing a million people, the damming of the holy Narmada River will submerge large areas of forest and farmland and create problems of waterlogging and salinization. Biodiversity is well-protected by a network of national parks and sanctuaries, as well as specific projects to conserve individual species, such as Operation Tiger. Urban areas are heavily polluted by industrial and vehicle emissions.*

---

indicator species plant or animal whose presence or absence in an area indicates certain environmental conditions, such as soil type, high levels of pollution, or, in rivers, low levels of dissolved oxygen. Many plants show a preference for either alkaline or acid soil conditions, while certain trees require aluminium, and are found only in soils where it is present. Some lichens are sensitive to sulphur dioxide in the air, and absence of these species indicates atmospheric pollution. The type of invertebrate found in a stream or river will indicate its pollution levels.

Indonesia *environmentalists have expressed concern over logging operations in the country's rainforest areas, both on grounds of damage to the ecosystem and the displacement of the indigenous population. Comparison of primary forest and 30-year-old secondary forest has shown that logging in Kalimantan has led to a 20% decline in tree species.*

indri largest living ⇨lemur *Indri indri* of Madagascar. Black and white, almost tailless, it has long arms and legs. It grows to 70 cm/2.3 ft long. It lives in trees and is active during the day. Its howl is doglike or human in tone. Like all lemurs, its survival is threatened by the widespread deforestation of Madagascar.

infiltration the passage of water into the soil. The rate of absorption of surface water by soil (the infiltration capacity) depends on the amount of surface water, the permeability and compactness of the soil, and the extent to which it is already saturated with water. Once in the soil, water may pass into the bedrock to form ⇨groundwater.

inorganic compound compound found in organisms that are not typically biological.

Water, sodium chloride, and potassium are inorganic compounds because they are widely found outside living cells. The term is also applied to those compounds that do not contain carbon and that are not manufactured by organisms. However, carbon dioxide is considered inorganic, contains carbon, and is manufactured by organisms during respiration. See ⇨organic compound.

insulation process or material that prevents or reduces the flow of electricity, heat or sound from one place to another.

U thermal or *heat insulation* makes use of insulating materials such as fibreglass to reduce the loss of heat through the roof and walls of buildings. The U-value of a material is a measure of its ability to conduct heat – a material chosen as an insulator should therefore have a low U-value. Air trapped between the fibres of clothes acts as a thermal insulator, preventing loss of body warmth.

Integrated Pest Management (IPM) the use of a coordinated array of methods to control pests, including biological control, chemical ⇨pesticides, ⇨crop rotation, and avoiding ⇨monoculture. By cutting back on the level of chemicals used the system can be both economical and beneficial to human health and the environment.

intermediate technology application of mechanics, electrical engineering, and other technologies based on designs developed in the developed world but utilizing materials, assembly and maintenance found in the developing world. See also ⇨appropriate technology.

Intermediate Technology Development Group UK-based international aid organization established 1965 by German economist E F Schumacher (1911–1977) to give advice and assistance on the appropriate choice of technologies for the rural poor of the Third World. It is an independent charity financed through donations.

Intermediate Technology concentrates mainly on renewable energy systems and small-scale manufacturing, and runs practical projects in many countries. The design and manufacture of a fuel-efficient stove and a bicycle trailer were among its 1980s successes.

internal-combustion engine heat engine in which fuel is burned inside the engine, contrasting with an external combustion engine (such as the steam engine) in which fuel is burned in a separate unit. The ⇨diesel engine and ⇨petrol engine are both ⇨internal-combustion engines. Gas ⇨turbines and jet and rocket engines are sometimes also considered to be internal-combustion engines because they burn their fuel inside their combustion chambers. It produces ⇨hydrocarbons, ⇨carbon monoxide, and nitrogen oxide as by-products. See ⇨catalytic converter.

International Atomic Energy Agency (IAEA) agency of the United Nations established 1957 to advise and assist member countries in the development and application of nuclear power, and to guard against its misuse. It has its headquarters in Vienna, and is responsible for research centres in Austria and Monaco, and the International

windpump

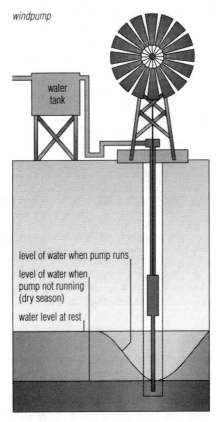

level of water when pump runs

level of water when
pump not running
(dry season)

water level at rest

intermediate technology *The simple windmill is an example of intermediate technology if it utilizes local materials and traditional design. In this way, there is no need for complex maintenance and repair, nor expensive spare parts.*

Centre for Theoretical Physics, Trieste, Italy, established 1964.

International Union for the Conservation of Nature organization established by the United Nations to promote the conservation of wildlife and habitats as part of the national policies of member states.

It has formulated guidelines and established research programmes (for example, International Biological Programme, IBP) and set up advisory bodies (such as Survival Commissions, SSC). In 1980, it launched the *World Conservation Strategy* to highlight particular problems, designating a small number of areas as *World Heritage Sites* to ensure their survival as unspoiled habitats (for example, Yosemite National Park in the USA, and the Simen Mountains in Ethiopia).

iodine greyish-black nonmetallic element, symbol I, atomic number 53, relative atomic mass 126.9044. It is a member of the halogen group. Its crystals give off, when heated, a violet vapour with an irritating odour resembling that of chlorine. It only occurs in combination with other elements. Its salts are known as iodides, which are found in sea water. As a mineral nutrient it is vital to the proper functioning of the thyroid gland, where it occurs in trace amounts as part of the hormone thyroxine. Iodine is used in photography, in medicine as an antiseptic, and in making dyes.

Its radioactive isotope ^{131}I (half-life of eight days) is a dangerous fission product from nuclear explosions and from the nuclear reactors in power plants, since, if ingested, it can be taken up by the thyroid and damage it.

ion exchange process whereby the ions in one compound replace the ions in another. The exchange occurs because one of the products is insoluble in water. For example, when hard water is passed over an ion-exchange resin, the dissolved calcium and magnesium ions are replaced by either sodium or hydrogen ions, so the hardness is removed. Commercial water softeners use ion-exchange resins. The addition of washing-soda crystals to hard water is also an example of ion exchange.

ionizing radiation radiation that knocks electrons from atoms during its passage, thereby leaving ions in its path. Alpha and beta particles are far more ionizing in their effect than are neutrons or gamma radiation.

Iraq the effects of the invasion of Kuwait and the Gulf war 1990–91 have caused severe environmental problems, mainly due to the damage done to the country's infrastructure by intensive bombing and subsequent interruption of food production. In particular, damage to water and sewage treatment plants has led to contamination of much of the country's water resources. A chemical-weapons plant covering an area of 65 sq km/25 sq mi, situated 80 km/50 mi NW of Baghdad, has been described by the UN as the largest toxic waste dump in the world.

irradiation method of preserving food by subjecting it to low-level ⇨gamma radiation in order to kill microorganisms. Although the process is now legal in several countries, uncertainty remains about possible long-term effects on consumers from irradiated food. The process does not make the food radioactive, but some vitamins, such as vitamin C, are destroyed and many molecular changes take place including the initiation of ⇨free radicals, which may be further changed into a range of unknown and unstable chemicals. Irradiation also eradicates the smell, taste, and poor appearance of bad or ageing food products.

irrigation artificial water supply for dry agricultural areas by means of dams and channels. Drawbacks are that it tends to concentrate salts, ultimately causing soil infertility, and that rich river silt is retained at dams, to the impoverishment of the land and fisheries below them.

Irrigation has been practised for thousands of years, in Eurasia as well as the Americas. An example is the channelling of the annual Nile flood in Egypt, which has been carried out from earliest times to its present control by the Aswan High Dam.

isotope one of two or more atoms that have the same atomic number (same number of protons), but which contain a different number of neutrons, thus differing in their atomic masses. They may be stable or radioactive, naturally occurring or synthesized.

Italy acidification from sulphur dioxide emissions by domestic and industrial sources is causing serious damage to the country's rich cultural heritage. Even artworks kept indoors can be damaged by airborne pollutants, as well as the thousands of important monuments and buildings across the country. Water pollution is also severe; most of the country's main lakes and rivers are so contaminated with industrial and agricultural effluent that they are no longer safe to swim in. Marine life along the Adriatic coast is under threat from chronic ⇨algal bloom.

IUCN abbreviation for ⇨International Union for the Conservation of Nature.

ivory the hard white substance of which the teeth and tusks of certain mammals are composed. Among the most valuable are elephants' tusks, which are of unusual hardness and density. Ivory is used in carving and other decorative work, and is so valuable that poachers continue to illegally destroy the remaining wild elephant herds in Africa to obtain it, despite heavy penalties and 'shoot-on-sight' policies in some countries.

Poaching for ivory has led to the decline of the African elephant population from 2 million to approximately 600,000, with the species virtually extinct in some countries. Trade in ivory was halted by Kenya 1989, but Zimbabwe continued its policy of controlled culling to enable the elephant population to thrive and to release ivory for export. China and Hong Kong have refused to obey an international ban on ivory trading.

J

jackal any of several wild dogs of the genus *Canis*, found in S Asia, S Europe, and N Africa. Jackals can grow to 80 cm/2.7 ft long, and have greyish-brown fur and a bushy tail. The Simien jackal, found in the Bale Mountains of Ethiopia, is close to extinction due to persecution by humans, increased agricultural activity, and its interbreeding with dogs and so is now the subject of conservation measures.

jaguar largest species of cat *Panthera onca* in the Americas, formerly ranging from the southwestern USA to southern South America, but now extinct in most of North America. It can grow up to 2.5 m/8 ft long including the tail. Formerly hunted for its beautiful coat, the jaguar is now protected by law in most countries, although it remains under threat from the destruction of its habitat and poaching.

JET (abbreviation for *Joint European Torus*) ⇨tokamak machine built in England to conduct experiments on nuclear fusion. It is the focus of the European effort to produce a practical fusion-power reactor. On 9 November 1991, the JET tokamak produced a 1.7 megawatt pulse of power in an experiment that lasted two seconds. This was the first time that a substantial amount of energy has been produced by nuclear power in a controlled experiment.

Jentink's duiker small (length 1.35 m/4.4 ft) and shy antelope *Cephalophus jentinki* that plunges into bushes when startled. It is acutely threatened by deforestation in its remaining habitat in W Africa where it is also hunted. One captive breeding colony exists in Texas and there are hopes of establishing others as the immediate future for this species in the wild appears to be bleak.

K

kakapo nocturnal, flightless parrot *Strigops habroptilus* that lives in burrows in New Zealand. It is green, yellow, and brown and weighs up to 3.5 kg/7.5 lb. When in danger, its main defence is to keep quite still. Because of the introduction of predators such as dogs, cats, rats, and ferrets, it is in danger of extinction as there are now only about 40 birds left.

Kazakhstan *nuclear testing during the Soviet era has resulted in much of the country being polluted by radiation. E Kazakhstan is estimated to have the highest level of ⇨background radiation in the world. Irrigation schemes which diverted waterways to the Aral Sea have had a devastating effect on fisheries and led to land degradation as surrounding land suffers from salination.*

Kenya *the elephant faces extinction as a result of poaching, although recently the government has adopted tough conservation measures to protect both elephant and rhino herds on which much of the country's income from tourism depends. Soil erosion is a serious problem in many areas as a result of overcultivation following migration to the fertile central plains. In an attempt to compensate many farmers have used too much fertilizer, further degrading the land.*

Kew Gardens popular name for the Royal Botanic Gardens, Kew, Surrey, England. They were founded 1759 by the mother of King George III as a small garden and passed to the nation by Queen Victoria 1840. By then they had expanded to almost their present size of 149 hectares and since 1841 have been open daily to the public. They contain a collection of over 25,000 living plant species and many fine buildings. The gardens are also a centre for botanical research.

Since 1964 there have been additional grounds at Wakehurst Place, Ardingly, West Sussex. The seeds of 5,000 species are preserved there in the seed physiology department, 2% of those known to exist in the world.

Kiribati *the islands are threatened by the possibility of a rise in sea level due to ⇨global warming. A rise of approximately 30 cm/1 ft by the year 2040 will make existing fresh water brackish and undrinkable*

Kuwait *during the Gulf War 1990–91, 650 oil wells were set alight and about 300,000 tonnes of oil were released into the waters of the Gulf killing thousands of birds and animals and leading to pollution haze, photochemical smog, acid rain, soil contamination, and water pollution. The latter is especially serious as Kuwait is extremely arid and lacks sufficient natural water resources to sustain agricultural activity, having to rely on water from ⇨aquifers.*

L

lake body of still water lying in depressed ground without direct communication with the sea. Lakes are common in formerly glaciated regions, along the courses of slow rivers, and in low land near the sea.

Agricultural fertilizers may leach into lakes from the land, causing ⇨eutrophication, an excessive enrichment of the water that causes an explosion of algal growth. This then depletes the lake's oxygen supply until it is no longer able to support life.

land set-aside scheme policy introduced by the European Community in the late 1980s, as part of the Common Agricultural Policy, to reduce overproduction of certain produce. Farmers are paid not to use land but to keep it fallow. The policy may bring environmental benefits by limiting the amount of fertilizers and pesticides used.

landfill site huge holes used for dumping household and commercial waste. Landfill disposal has been the preferred option in the UK and the USA for many years, with up to 85% of household waste being dumped in this fashion. However, the sites can be dangerous, releasing toxins and other leachates into the soil and the policy is itself wasteful both in terms of the materials dumped and land usage. Decomposing organic matter releases methane, which can be explosive, although many sites collect the gas and burn it for energy.

Many of the items found in landfill sites, for instance bottles, tins and cans, will remain intact for hundreds of years, and would be better reused or recycled. The UK government has recommended that local councils should aim to recycle 25% of all household rubbish. In principle over 50% of such waste is recyclable. At present in the UK only 5% of household waste is recovered.

Large Electron Positron Collider (LEP) world's largest particle accelerator, in operation from 1989 at the CERN laboratories near Geneva in Switzerland. It occupies a tunnel 3.8 m/12.5 ft wide and 27 km/16.7 mi long, which is buried 180 m/590 ft underground and forms a ring consisting of eight curved and eight straight sections. In 1989 the LEP was used to measure the mass and lifetime of the Z particle, carrier of the weak nuclear force.

Electrons and positrons enter the ring after passing through the Super Proton Synchrotron accelerator. They travel in opposite directions around the ring, guided by 3,328 bending magnets and kept within tight beams by 1,272 focusing magnets. As they pass through the straight sections, the particles are accelerated by a pulse of radio energy. Once sufficient energy is accumulated, the beams are allowed to collide. Four giant detectors are used to study the resulting shower of particles.

leaching process by which substances are washed out of the soil. Fertilizers leached out of the soil drain into rivers, lakes, and ponds and cause water pollution. The risk of nitrates and pesticides leaching into ⇨aquifers is greatest in sandy or shallow soils. In tropical areas, leaching of the soil after the destruction of forests removes scarce nutrients and can lead to a dramatic loss of soil fertility. The leaching of soluble minerals in soils can lead to the formation of distinct soil horizons as different minerals are deposited at successively lower levels.

lead heavy, soft, malleable, grey, metallic element, symbol Pb, atomic number 82, relative atomic mass 207.19. Usually found as an ore (most often in galena), it occasionally occurs as a free metal (native metal), and is the final stable product of the decay of uranium. Lead is the softest and weakest of the commonly used metals, with a low melting point; it is a poor conductor of electricity and resists acid corrosion. As a cumulative poison, lead enters the body from lead water pipes, lead-based paints, and leaded petrol. In humans, exposure to lead shortly after birth is associated with impaired mental health between the ages of two and four. The metal is an effective shield against radiation and is used in batteries, glass, ceramics, and alloys such as pewter and solder.

leaded petrol petrol that contains antiknock, a mixture of the chemicals tetraethyl lead and dibromoethane. The lead from the exhaust fumes enters the atmosphere, mostly as simple lead compounds, and is poisonous to children's developing neural systems. Leaded petrol destroys ⇨catalytic converters.

Lebanon *the recurrent civil war since 1976 has caused massive damage to the infrastructure and environment. Urbanization has resulted in the loss of much of the country's agricultural land and wildlife habitat. In the coastal areas, industrial effluent and other waste has polluted large stretches of the Mediterranean. Intensive fishing techniques (such as the use of dynamite to stun whole shoals of fish which can then be collected at leisure) have also caused serious damage to the marine ecosystem.*

lemur prosimian primate of the family Lemuridae, inhabiting Madagascar and the Comoros Islands. There are about 16 species, ranging from mouse-sized to dog-sized animals. Lemurs are tree-living, and some species are nocturnal. They have long, bushy tails, and feed on fruit, insects, and small animals. Many are threatened with extinction owing to loss of their forest habitat and, in some cases, from hunting.

leopard or *panther* cat *Panthera pardus*, found in Africa and Asia. The background colour of the coat is golden, and the black spots form rosettes that differ according to the variety; black panthers are simply a colour variation and retain the patterning as a 'watered-silk' effect. The leopard is 1.5–2.5 m/5–8 ft long, including the tail, which may measure 1 m/3 ft.

Libya *shortage of water is a serious problem due to the arid climate, exacerbated by the absence of rivers. A plan to pump water from below the Sahara to the coast (Great Manmade River Project) risks rapid exhaustion of a nonrenewable supply as well as possible contamination of ⇨aquifers. Partly as a result of the problems with water supply, Libya also suffers from land degradation and desertification. Waste and industrial effluents add to the pollution of the Mediterranean.*

lichen plant-like growth on rocks and trees, actually an association of a fungus and ⇨algae. They grow slowly, obtaining nutrition through the photosynthetic activities of the algae, which live inside the fungus, benefiting from shelter and from a supply of minerals, water, and carbon dioxide. Lichens vary in form, being crusty, leafy, or bushy. They are reliable ⇨indicator species because they are killed by sulphur dioxide pollution, particularly the bushy varieties, which can only be found where sulphur dioxide levels are negligible.

life cycle the sequence of developmental stages through which members of a given species pass. Most vertebrates have a simple life cycle consisting of fertilization of sex cells or gametes, a period of development as an embryo, a period of juvenile growth after hatching or birth, an adulthood including sexual reproduction, and finally death. Invertebrate life cycles are generally more complex and may involve major reconstitution of the individual's appearance (metamorphosis) and completely different styles of life. Plants have a special type of life cycle with two distinct phases, known as alternation of generations. Many insects such as cicadas, dragonflies, and mayflies have a long larvae or pupae phase and a short adult phase. Dragonflies live an aquatic life as larvae and an aerial life during the adult phase. In many invertebrates and protozoa there is a sequence of stages in the life cycle, and in parasites different stages often occur in different host organisms.

life-cycle analysis assessment of the environmental impact of a product, taking into account all aspects of production (including resources used), packaging, distribution and ultimate end. This 'cradle-to-grave' approach can expose the fallacy of many so-called 'eco-friendly' labels, applied to products such as soap powders, which may be biodegradable but which are perhaps contained in nonrecyclable containers.

life expectancy average lifespan of a person at birth. It depends on many factors such as nutrition, disease control, environmental contaminants, war, stress, and general living standards.

There is a marked difference between industrialized countries, which generally have an ageing population, and the poorest countries, where life expectancy is much shorter. In Bangladesh, life expectancy is currently 48; in Nigeria 49; in famine-prone Ethiopia it is only 41; whereas in the UK, average life expectancy for both sexes currently stands at 75 and heart disease is the main cause of death.

linear accelerator or *linac* a type of particle accelerator in which the particles move along a straight tube. Particles pass through a linear accelerator only once – unlike those in a cyclotron (a ring-shaped accelerator), which make many revolutions, gaining energy each time.

The world's longest linac is the Stanford Linear Collider, in which electrons and positrons are accelerated along a straight track 3.2 km/ 2 mi long and then steered into a head-on collision with other particles.

Lithuania *the country has no natural fuel resources and so is entirely dependent on imported fuel and energy supplied by Ignalina nuclear power plant, built to the same design as the ⇨Chernobyl reactor. Industrial pollution is a serious concern, with many workers overexposed to toxic substances and high levels of air pollution in urban areas. However, since independence from the former Soviet Union 1991, the Lithuanian government has introduced a series of laws the and energy conservation programmes to bring the country into line with EC environmental standards.*

luminous paint preparation containing a mixture of pigment, oil, and a phosphorescent sulphide, usually calcium or barium. After exposure to light it appears luminous in the dark. The luminous paint used on watch faces contains radium, is radioactive and therefore does not require exposure to light. The qauntity of radiation emitted is negligible and so does not provide a health risk.

M

Madagascar *according to 1990 UN figures, 93% of the forest area has been destroyed and about 100,000 species have been made extinct partly as a result. Only one in ten of the population has access to safe drinking water in rural areas.*

magnetohydrodynamics (MHD) field of science concerned with the behaviour of ionized gases or liquid in a magnetic field. Systems have been developed that use MHD to generate electrical power.

MHD-driven ships have been tested in Japan. In 1991 two cylindrical thrusters with electrodes and niobium–titanium superconducting coils, soaked in liquid helium, were placed under the passenger boat *Yamato 1*. The boat, 30 m/100 ft long, was designed to travel at 8 knots. An electric current passed through the electrodes accelerates water through the thrusters, like air through a jet engine, propelling the boat forward.

Magnox early type of nuclear reactor used in the UK, for example in Calder Hall, the world's first commercial nuclear power station. This type of reactor uses uranium fuel encased in tubes of magnesium alloy called Magnox. Carbon dioxide gas is used as a coolant to extract heat from the reactor core. See also ⇨nuclear energy.

mahogany timber from any of several genera of trees found in the Americas and Africa. Mahogany is a tropical hardwood obtained chiefly by rainforest logging. It has a warm red colour and takes a high polish.

True mahogany comes from trees of the genus *Swietenia*, but other types come from the Spanish and Australian cedars, the Indian redwood, and other trees of the mahogany family Meliaceae, native to Africa and the E Indies. The species is under threat due to its popularity in the West for use in the manufacture of musical instruments, furniture, and veneers. There are attempts to make it the first hardwood tree listed under the CITES convention and recently hardwood suppliers, DIY shops, and importers have been the subject of protests and boycott campaigns by environmentalists.

Mali *a rising population coupled with recent droughts has affected marginal agriculture. Once in surplus, Mali has had to import grain every year since 1965. The same factors, together with poaching, have had a disastrous effect on the country's formerly rich species diversity; if the current rate of decline continues there will be no significant mammal populations within a decade.*

Malthus theory projection of population growth made by English economist Thomas Malthus in 1793. He based his theory on the ⇨population explosion that was already becoming evident in the late 18th century, and argued that the number of people would increase faster than the food supply. Population would eventually reach a resource limit (⇨overpopulation). Any further increase would result in a population crash, caused by famine, disease, or war.

Malthus was not optimistic about the outcome and suggested that only 'moral restraint' (birth control) could prevent crisis. Malthus is criticized by some economists who argue that it is population growth which causes development, as productivity must be increased and so new technologies and agricultural techniques are developed. Under these theories, there is no direct causal link between population and famine.

mangrove swamp muddy swamp found on tropical coasts and estuaries, characterized by dense thickets of mangrove trees. These low trees are adapted to live in creeks of salt water and send down special breathing roots from their branches to take in oxygen from the air. The roots trap silt and mud, creating a firmer drier environment over time. Mangrove swamps are common in the Amazon delta and along the coasts of W Africa, N Australia, and Florida, USA.

meltdown the melting of the core of a nuclear reactor, due to overheating. To prevent such accidents all reactors have equipment intended to flood the core with water in an emergency. The reactor is housed in a strong containment vessel, designed to prevent radiation escaping into the atmosphere. The result of a meltdown is an area radioactively contaminated for 25,000 years or more.

At Three Mile Island, Pennsylvania, USA, in March 1979, a partial meltdown occurred caused by a combination of equipment failure and operator error, and some radiation was released into the air. In April 1986, a reactor at ⇨Chernobyl, near Kiev, Ukraine, exploded, causing a partial meltdown of the core. Radioactive ⇨fallout was detected as far away as Canada and Japan.

Mexico *Air is polluted by 130,000 factories and 2.5 million vehicles; during the 1980s, smog levels in Mexico City exceeded World Health Organization standards on more than 300 days of the year, although recently measures have been taken to curb vehicle emissions. Water and land pollution are also severe, especially in the heavily industrialized zones on the US border.*

Minamata Japanese town where 43 people died 1953–56 after eating fish poisoned with dimethyl mercury. The poison had been released as ⇨effluent from a local plastics factory, and became concentrated in the flesh of sea organisms. Many townspeople suffered long-terms effects, including paralysis, tremors, and brain damage.

mixed farming farming system where both arable and pastoral farming is carried out. Mixed farming is a lower-risk strategy than ⇨monoculture. If climate, pests, or market prices are unfavourable for one crop or type of livestock, another may be more successful and the risk is shared. Animals provide manure for the fields and help to maintain soil fertility.

moderator in a thermal reactor, a material such as graphite or heavy water used to reduce the speed of high-energy neutrons.

monoculture farming system where only one crop is grown. In developing countries this is often a ⇨cash crop, grown on plantations. Cereal crops in the industrialized world are also frequently grown on a monoculture basis; for example, wheat in the Canadian prairies.

Monoculture allows the farmer to tailor production methods to the requirements of one crop, but it is a high-risk strategy since the crop may fail (because of pests, disease, or bad weather) or world prices for the crop may fall. Monoculture without ⇨crop rotation is likely to result in reduced soil quality despite the addition of artificial fertilizers, and it contributes to ⇨soil erosion.

Montréal Protocol international agreement, signed 1987, to stop the production of chemicals that are ⇨ozone depleters by the year 2000.

Originally the agreement was to reduce the production of ozone depleters by 35% by 1999. The green movement criticized the agreement as inadequate, arguing that an 85% reduction in ozone depleters would be necessary just to stabilize the ozone layer at 1987 levels. The protocol (under the Vienna Convention for the Protection of the Ozone Layer) was reviewed 1992. Amendments added another 11 chemicals to the original list of eight chemicals suspected of harming the ozone layer. A controversial amendment concerns a fund established to pay for the transfer of ozone-safe technology to poor countries.

mountain gorilla highly endangered ape subspecies *Gorilla gorilla beringei* found in rainforest on the Rwanda, Zaire, and Uganda borders in central Africa, with a total population of under 400. It is threatened by deforestation and illegal hunting for skins and the zoo trade.

Mozambique the country was devastated during the 1980s by a lethal cycle of drought, war, and famine which combined to make epidemics (spread quickly through large refugee camps) and famine the major cause of death apart from the intermittent civil war. Infant mortality is particularly high, although as the situation stabilizes the picture is marginally improving.

mutagen any substance that makes mutation of genes more likely. A mutagen is likely to also act as a ⇨carcinogen.

Myanmar illegal but largely unchecked logging and fires have resulted in the loss of over two-thirds of the country's rainforest. Landslides and flooding during the rainy season (June–Sept) are becoming more frequent as a result of deforestation. Political unrest and repression have diverted attention so that there is no serious effort to tackle the country's environmental problems.

N

national park land set aside and conserved for public enjoyment. The first was Yellowstone National Park, USA, established 1872. National parks include not only the most scenic places, but also places distinguished for their historic, prehistoric, or scientific interest, or for their superior recreational assets. They range from areas the size of small countries to pockets of just a few hectares.

In England and Wales under the National Park Act 1949 10 national parks were established, including the Peak District, Lake District, and Snowdonia. These cover some 1.4 million hectares or 9% of the total land area of England and Wales. Some countries have ⇨wilderness areas, with no motorized traffic, no overflying aircraft, no hotels, hostels, shops, or cafés, no industry, and the minimum of management. In the UK national parks are not wholly wilderness or conservation areas, but merely places where planning controls on development are stricter than elsewhere. However, from time to time pressure to develop land for agriculture, quarrying, or tourism, or to improve amenities for the local community means that conflicts of interest arise between land-users. Other protected areas include Areas of Outstanding Natural Beauty, ⇨Sites of Special Scientific Interest (SSSIs), and National Scenic Areas.

National Rivers Authority (NRA) UK government agency launched 1989. It is responsible for managing water resources, investigating and regulating pollution, and taking over flood controls and land drainage from the former 10 regional water authorities of England and Wales. It also has responsibilities for salmon and freshwater fisheries, navigation in some regions, and nature conservation and recreation in inland waters.

Natural Environment Research Council (NERC) UK organization established by royal charter 1965 to undertake and support research in the earth sciences, to give advice both on exploiting natural resources and on protecting the environment, and to support education and training of scientists in these fields of study.

Research areas include geothermal energy, industrial pollution, waste disposal, satellite surveying, acid rain, biotechnology, atmospheric circulation, and climate. Research is carried out principally within the UK but also in Antarctica and in many Third World countries. It comprises 13 research bodies.

Within the NERC, the research bodies are: Freshwater Biological Association, British Geological Survey, Institute of Hydrology, Plymouth Marine Laboratory, Institute of Oceanographic Sciences, Deacon Laboratory, Proudman Oceanographic Laboratory, Institute of Virology, Institute of Terrestrial Ecology, Scottish Marine Biological Association, Sea Mammal Research Unit, Unit of Comparative Plant Ecology, and the NERC Unit for Thematic Information Systems.

natural gas mixture of flammable gases found in the Earth's crust (often in association with petroleum). It is one of the world's three main fossil fuels (with coal and oil). Natural gas is a mixture of ⇨hydrocarbons, chiefly methane, with ethane, butane, and propane. Natural gas is usually transported from its source by pipeline, although it may be liquefied for transport and storage and is, therefore, often used in remote areas where other fuels are scarce and expensive. Prior to transportation, butane and propane are removed and liquefied to form 'bottled gas'.

Test flights of the first aircraft powered by liquefied natural gas began 1989. The Soviet-produced craft will save 9 tonnes of kerosene on a journey of 2,000 km/1,250 mi.

In the UK from the 1970s natural gas from the North Sea has superseded coal gas, or town gas, both as a domestic fuel and as an energy source for power stations.

natural radioactivity radioactivity generated by those radioactive elements that exist in the Earth's crust. All the elements from polonium (atomic number 84) to uranium (atomic number 92) are radioactive. ⇨Radioisotopes of some lighter elements are also found in nature (for example potassium-40). Natural radioactivity comprises 87% of the total annual average dose. The most important source of natural radiation is the gas radon, which forms in the ground and seeps into the atmosphere.

nature the living world, including plants, animals, fungi, and all microorganisms, and naturally formed features of the landscape, such as mountains and rivers.

Nature Conservancy Council (NCC) former name of UK government agency divided 1991 into English Nature, Scottish Natural Heritage, and the Countryside Council for Wales.

The NCC was established by act of Parliament 1973 (Nature Conservancy created by royal charter 1949) with the aims of designating and managing national nature reserves and other conservation areas; identifying Sites of Special Scientific Interest; advising government ministers on policies; providing advice and information; and commissioning or undertaking relevant scientific research. In 1991 the Nature Conservancy Council was dissolved and its three regional bodies became autonomous agencies.

nature reserve area set aside to protect a habitat and the wildlife that lives within it, with only restricted admission for the public. A nature reserve often provides a sanctuary for rare species. The world's largest is Etosha Reserve, Namibia; area 99,520 sq km/38,415 sq mi.

In Britain, there are both officially designated nature reserves – managed by English Nature, the Countryside Council for Wales, and Scottish Natural Heritage – and those run by a variety of voluntary conservation organizations.

neo-Darwinism modern theory of evolution, built up since the 1930s by integrating the 19th-century English scientist Charles Darwin's theory of evolution through natural selection with the theory of genetic inheritance founded on the work of the Austrian biologist Gregor Mendel.

Neo-Darwinism asserts that evolution takes place because the environment is slowly changing, exerting a selection pressure on the individuals within a population. Those with characteristics that happen to adapt to the new environment are more likely to survive and have offspring and hence pass on these favourable characteristics. Over time the genetic make-up of the population changes and ultimately a new species is formed.

Nepal *tourism is a major environmental problem, but also a vital source of income. Described as the world's highest rubbish dump, Nepal attracts 270,000 tourists, trekkers, and mountaineers each year. An estimated 500 kg/1,100 lb of rubbish is left by each expedition trekking or climbing in the Himalayas. Since 1952 the foothills of the Himalayas have been stripped of 40% of their forest cover for fuel and cultivation; deforestation is a problem nationwide and there is little effort to replace the forest lost. These problems all contribute to land degradation, making the pressure to cultivate fresh land even more intense.*

NERC abbreviation for ⇨Natural Environment Research Council.

Netherlands *the country lies at the mouths of three of Europe's most polluted rivers, the Maas, Rhine, and Scheldt. Dutch farmers contribute to this pollution by using the world's highest concentrations of nitrogen-based fertilizer per hectare per year. Despite this, the country has a fairly good environmental record, with the world's only coherent policy for creating a sustainable society and an effective system for recycling domestic waste.*

neutral solution solution of pH 7, in which the concentrations of hydrogen ($H^+_{(aq)}$) and hydroxide ($OH^-_{(aq)}$) ions are equal.

neutron one of the three main subatomic particles, the others being the proton and the electron. The neutron is a composite particle, being made up of three quarks, and therefore belongs to the baryon group of the hadrons. Neutrons have about the same mass as protons but no electric charge, and occur in the nuclei of all atoms except hydrogen. They contribute to the mass of atoms but do not affect their chemistry.

For instance, the ⇨isotopes of a single element differ only in the number of neutrons in their nuclei but have identical chemical properties. Outside a nucleus, a free neutron is radioactive, decaying with a half-life of 11.6 minutes into a proton, an electron, and an antineutrino.

neutron bomb small hydrogen bomb for battlefield use that kills by radiation without destroying buildings and other structures.

Nicaragua *over a decade of civil war has prevented attention being focused on the environment, but the situation has now largely stabilized. Deforestation in the west of the country has created chronic soil erosion and the heavy use of pesticides since the early 1970s has contaminated the water supply.*

niche the 'place' occupied by a species in its habitat, including all chemical, physical, and biological components, such as what it eats, the time of day at which the species feeds, temperature, moisture, the parts of the habitat that it uses (for example, trees or open grassland), the way it reproduces, and how it behaves.

It is believed that no two species can occupy exactly the same niche, because they would be in direct competition for the same resources at every stage of their life cycle.

nitrate salt or ester of nitric acid, containing the NO_3^- ion. Nitrates are used in explosives, in the chemical and pharmaceutical industries, in curing meat, and as fertilizers. They are the most water-soluble salts known and play a major part in the ⇨nitrogen cycle. Nitrates in the soil, whether naturally occurring or from inorganic or organic fertilizers, can be used by plants to make proteins and nucleic acids. However, run-off from fields can result in ⇨nitrate pollution.

nitrate pollution the contamination of water by nitrates. Increased use of artificial fertilizers and land cultivation means that higher levels of nitrates are being washed from the soil into rivers, lakes, and ⇨aquifers. There they cause an excessive enrichment of the water (⇨eutrophication), leading to a rapid growth of algae, which in turn darkens the water and reduces its oxygen content. The water is expensive to purify and many plants and animals die. High levels are now found in drinking water in arable areas. These may be harmful to newborn babies, and it is possible that they contribute to stomach cancer, although the evidence for this is unproven.

The UK current standard is 100 milligrams per litre. This is double the EC limits implemented in 1993.

nitrification process that takes place in soil when bacteria oxidize ammonia, turning it into nitrates. Nitrates can be absorbed by the roots of plants, so this is a vital stage in the ⇨nitrogen cycle.

nitrite salt or ester of nitrous acid, containing the nitrite ion (NO_2^-). Nitrites are used as preservatives (for example, to prevent the growth of botulism spores) and as colouring agents in cured meats such as bacon and sausages.

nitrogen colourless, odourless, tasteless, gaseous, nonmetallic element, symbol N, atomic number 7, relative atomic mass 14.0067. It forms almost 80% of the Earth's atmosphere by volume and is a constituent of all plant and animal tissues (in proteins and nucleic acids). Nitrogen is obtained for industrial use by the liquefaction and fractional distillation of air. Its compounds are used in the manufacture of foods, drugs, fertilizers, dyes, and explosives.

Nitrogen has been recognized as a plant nutrient, found in manures and other organic matter, from early times, long before the complex cycle of ⇨nitrogen fixation was understood.

nitrogen cycle the process of nitrogen passing through the ecosystem. Nitrogen, in the form of inorganic compounds (such as nitrates) in the soil, is absorbed by plants and turned into organic compounds (such as proteins) in plant tissue. A proportion of this nitrogen is eaten by herbivores, with some of this in turn being passed on to the carnivores, which feed on the herbivores. The nitrogen is ultimately returned to the soil as excrement and when organisms die and are converted back to inorganic form by ⇨decomposers.

Although about 78% of the atmosphere is nitrogen, this cannot be used directly by most organisms. However, certain bacteria and cyanobacteria (see ⇨blue-green algae) are capable of nitrogen fixation. Some nitrogen-fixing bacteria live mutually with leguminous plants (peas and beans) or other plants (for example, alder), where they form characteristic nodules on the roots. The presence of such plants increases the nitrate content, and hence the fertility, of the soil.

nitrogen fixation the process by which nitrogen in the atmosphere is converted into nitrogenous compounds (such as nitrates) by the action of microorganisms, such as cyanobacteria (see ⇨blue-green algae) and bacteria, in conjunction with certain legumes in the ⇨nitrogen cycle. Several chemical processes duplicate this to produce fertilizers.

noise unwanted sound. Permanent, incurable loss of hearing can be caused by prolonged exposure to high noise levels (above 85 decibels).

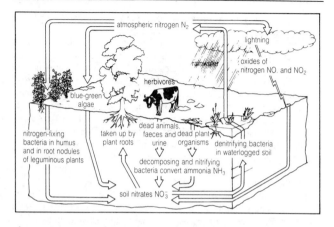

nitrogen cycle The nitrogen cycle is one of a number of cycles during which the chemicals necessary for life are recycled. The carbon, sulphur, and phosphorus cycles are others. Since there is only a limited amount of these chemicals in the Earth and its atmosphere, the chemicals must be continuously recycled if life is to go on.

Noise is a recognized form of pollution, but is difficult to actually measure, because the annoyance or discomfort caused varies between individuals. The World Health Organization regards noise over 55 decibels – roughly the noise level of a conversation – on a daily outdoor basis as an unacceptable level. It is estimated 130 million people were exposed to noise of this level or greater in 1991.

If the noise is in a narrow frequency band, temporary hearing loss can occur even though the level is below 85 decibels or exposure is only for short periods. Lower levels of noise are an irritant, but seem not to increase fatigue or affect efficiency to any great extent. Roadside meter tests, introduced by the Ministry of Transport in Britain in 1968, allowed 87 decibels as the permitted limit for cars and 92 for lorries. Loud noise is a major pollutant in towns and cities. In the UK the worst source of noise nuisance is noise from neighbours, suffered by 20% of the population.

Loss of hearing is a common complaint of people working on factory production lines or in the construction and road industry. Minor psychiatric disease, stress-related ailments including high blood pressure, and disturbed sleep patterns are regularly linked to noise, although the causal links are in most cases hard to establish. In the UK, noise control is the responsibility of the Local Authorities, whose Environmental Health Officers operate under powers devolved to them through the Control of Pollution Act 1974, the Environment Protection Act 1990, and the Town and Country Planning Act 1990. Noise in the workplace, and noise generated by airports, is controlled by separate legislation. It is estimated that 60% of all noise complaints relate to domestic property, dogs and hi-fi systems being the major preoccupation.

nonrenewable resource natural resource, such as coal or oil, that takes thousands or millions of years to form naturally and can therefore not be replaced once it is consumed. The main energy sources used by humans are nonrenewable; ⇨renewable resources, such as solar, tidal, and geothermal power, have so far been less exploited.

Nonrenewable resources have a high carbon content because their origin lies in the photosynthetic activity of plants millions of years ago. The fuels release this carbon back into the atmosphere as carbon dioxide. The rate at which such fuels are being burnt is thus resulting in a rise in the concentration of carbon dioxide in the atmosphere.

Norway *an estimated 80% of the lakes and streams in the southern half of the country have been severely acidified by acid rain. Despite efforts to reduce their own emissions of sulphur dioxide (by 60% over the last decade) the Norwegians are hampered by the fact that over 90% of the acid rain is caused by airborne pollutants from other European countries.*

---

nuclear energy or *atomic energy* energy released from the nucleus of the atom. Energy produced by *nuclear fission* (the splitting of uranium or plutonium nuclei) has been harnessed since the 1950s to generate electricity, and research continues into the possible controlled use of ⇨nuclear fusion (the fusing, or combining, of atomic nuclei).

nuclear energy: chronology

1896	French physicist Henri Becquerel discovered radioactivity.
1905	In Switzerland, Albert Einstein showed that mass can be converted into energy.
1911	New Zealand physicist Ernest Rutherford proposed the nuclear model of the atom.
1919	Rutherford split the atom, by bombarding a nitrogen nucleus with alpha particles.
1939	Otto Hahn, Fritz Strassman, and Lise Meitner announced the discovery of nuclear fission.
1942	Enrico Fermi built the first nuclear reactor, in a squash court at the University of Chicago, USA.
1946	The first fast reactor, called Clementine, was built at Los Alamos, New Mexico.
1951	The Experimental Breeder Reactor, Idaho, USA, produced the first electricity to be generated by nuclear energy.
1954	The first reactor for generating electricity was built in the USSR, at Obninsk.
1956	The world's first commercial nuclear power station, Calder Hall, came into operation in the UK.
1957	The release of radiation from Windscale (now Sellafield) nuclear power station, Cumbria, England, caused 39 deaths to 1977. In Kyshym, USSR, the venting of plutonium waste caused high but undisclosed casualties (30 small communities were deleted from maps produced in 1958).
1959	Experimental fast reactor built in Dounreay, N Scotland.
1979	Nuclear-reactor accident at Three Mile Island, Pennsylvania, USA.
1986	An explosive leak from a reactor at Chernobyl, the Ukraine, resulted in clouds of radioactive material spreading as far as Sweden; 31 people were killed and thousands of square kilometres were contaminated.
1991	The first controlled and substantial production of nuclear-fusion energy (a two-second pulse of 1.7 MW) was achieved at JET, the Joint European Torus, at Culham, Oxfordshire, England.
1993	The Tokamak Fusion Test Reactor at Princeton University, USA, generated 6.2 MW of power for four seconds – the most energy ever generated by controlled nuclear fusion.

In nuclear power stations, fission takes place in a ⇨nuclear reactor. The nuclei of uranium or, more rarely, plutonium are induced to split, releasing large amounts of heat energy. The heat is then removed from the core of the reactor by circulating gas or water, and used to produce the steam that drives alternators and turbines to generate electrical power.

Unlike fossil fuels, such as coal and oil, which must be burned in large quantities to produce energy, nuclear fuels are used in very small amounts and supplies are therefore unlikely to be exhausted in the foreseeable future. However, the use of nuclear energy has given rise to concern over safety. Anxiety has been heightened by accidents such as the one at Chernobyl, Ukraine, in 1986. There has also been mounting concern about the production and disposal of toxic nuclear waste, which may have an active life of several thousand years, and the cost of maintaining nuclear power stations and decommissioning them at the end of their lives.

nuclear fusion process whereby two atomic nuclei are fused, with the release of a large amount of energy. Very high temperatures and pressures are thought to be required in order for the process to happen. Under these conditions the atoms involved are stripped of all their electrons so that the remaining particles, which together make up a *plasma*, can come close together at very high speeds and overcome the mutual repulsion of the positive charges on the atomic nuclei. At very close range another nuclear force will come into play, fusing the particles together to form a larger nucleus. As fusion is accompanied by the release of large amounts of energy, the process might one day be harnessed to form the basis of commercial energy production. Methods of achieving controlled fusion are therefore the subject of research around the world.

Fusion is the process by which the Sun and the other stars produce their energy.

nuclear reactor central component of a nuclear power station that generates ⇨nuclear energy under controlled conditions for use as a source of electrical power. The nuclei of uranium-235 atoms undergo induced nuclear fission in the reactor, and release energy in many forms, one of which is heat. The heat is removed from the core of the reactor by circulating gas or water, and is used to produce the steam that, under high pressure, drives turbines and alternators to produce electrical power. See also ⇨advanced gas-cooled reactor and ⇨fast reactor.

nuclear safety measures to avoid accidents in the operation of nuclear reactors and in the production and disposal of nuclear weapons and of ⇨nuclear waste. There are no guarantees of the safety of any of the various methods of disposal.

nuclear accidents

Tomsk, Siberia. In April 1993 a tank exploded at a uranium reprocessing plant, sending a cloud of radioactive particles into the air.

Chernobyl, Ukraine. In April 1986 there was an explosive leak, caused by overheating, from a nonpressurized boiling-water reactor, one of the largest in Europe. The resulting clouds of radioactive material spread as far as the UK; 31 people were killed in the explosion (many more are expected to die or become ill because of the long-term effects of radiation), and thousands of square kilometres of land were contaminated by fallout.

Three Mile Island, Harrisburg, Pennsylvania, USA. In 1979, a combination of mechanical and electrical failure, as well as operator error, caused a pressurized water reactor to leak radioactive matter.

Church Rock, New Mexico, USA. In July 1979, 380 million litres/100 million gallons of radioactive water containing uranium leaked from a pond into the Rio Purco, causing the water to become over 6,500 times as radioactive as safety standards allow for drinking water.

Ticonderoga, 130 km/80 mi off the coast of Japan. In 1965 a US Navy Skyhawk jet bomber fell off the deck of this ship, sinking in 4,900 m/16,000 ft of water. It carried a one-megaton hydrogen bomb. The accident was only revealed in 1989.

Windscale (now Sellafield), Cumbria, England. In 1957, fire destroyed the core of a reactor, releasing large quantities of radioactive fumes into the atmosphere.

In 1990 a scientific study revealed an increased risk of leukaemia in children whose fathers had worked at Sellafield between 1950 and 1985. Sellafield (UK) is the world's greatest discharger of radioactive waste, followed by Hanford, Washington (USA).

nuclear waste the radioactive and toxic by-products of the nuclear-energy and nuclear-weapons industries. It presents a major disposal problem with no obvious solution. Increasingly the issue of nuclear waste is becoming the central controversy threatening the future of gen-

NUCLEAR WASTE

Although it is now almost 50 years since scientists achieved controlled nuclear fission, the question of what to do about nuclear waste is as controversial as ever. Nuclear waste comes from nuclear power stations, nuclear weapons, uranium mines, and waste from medical and industrial uses of nuclear energy. It is one of the most hazardous of all waste products, because it releases dangerous forms of radiation, capable of causing a range of immediate and long-term health effects including radiation sickness, cancer, and birth defects. Some nuclear waste products can stay dangerously radioactive for many thousands of years, thus creating a disposal problem of a type and scale not met previously.

Nuclear waste can be categorized into three main types: high-, medium-, and low-level waste. The spent uranium rods from nuclear power stations are among the highest-level waste products, as are nuclear-weapon components. Less radioactive waste includes cooling water used in power stations and medical equipment.

Early innocence

Radiation in nuclear waste cannot be destroyed by burning or industrial processes, and people regulating nuclear waste have a choice of storing it indefinitely, releasing it into the environment in the hope that it will disperse to nontoxic levels, or, in the case of some high-level wastes, reprocessing it to remove usable fuel. In practice, all three options are used. When nuclear power was first used, far less was known about its hazards, and large amounts of low-level waste were simply discharged into the environment, ending up in the air, in rivers and oceans, and sometimes in the soil. Since then, there has been a steady decrease in 'acceptable' levels of radiation from the point of view of health and a consequent tightening of controls.

Unfortunately, a lot of damage has already been done. Despite some changes in regulations, low-level waste is still lost by accident or design. Evidence collected by the environmental organization Greenpeace confirmed suspicions that the Soviet navy used to dump reactors from nuclear submarines in Arctic seas, some of them with fuel still inside. Clusters of child leukaemias around some of the UK's nuclear facilities have caused a storm of controversy among researchers, with some believing that they point to a link with

discharges of radiation, and others saying that the statistical proba-
bility of a link with the nuclear-power programme remains tenuous.

Storing nuclear waste

Many nuclear operations now choose to store their waste, until either
the radioactivity has fallen to background levels or new disposal
techniques are developed. The lowest-level waste is buried in metal
containers in shallow trenches. Some countries have built concrete
storage tanks for low- and intermediate-level waste, such as that in
Drigg, Cumbria, in the UK. High-level waste is sometimes stored
underwater until the most intense radioactivity has dissipated.

Options for long-term storage include burial deep underground,
with the waste vitrified (encased in glass), deposited in deep-sea
trenches, or buried in polar icecaps. Some nuclear scientists even
suggest blasting it off into space by rocket. None of these solutions
is without serious problems. Space disposal would be extremely
expensive and hazardous in case of accident. Sea dumping is banned
by international treaty. Burial underground is the favoured option at
present but, given the enormous time scale involved, it is impossible
for us to be sure that containers would not corrode, be displaced by
earth movements, and otherwise leak their contents. Simply transporting nuclear waste risks escapes of radioactive material.

Reprocessing: the most controversial option of all

The third disposal option, for high-level waste from spent fuel,
includes reprocessing to extract remaining uranium and plutonium.
The latter can be used to make nuclear weapons or as nuclear fuel.
Reprocessing has long been favoured by the nuclear industry, partly
because of greater fuel-use efficiency and partly because the close
links between civilian and military nuclear programmes in many
countries makes plutonium production an attractive option.

One of the best known reprocessing facilities is Sellafield in Cumbria, which processes nuclear waste from many parts of the world. It
suffered a serious accident in 1957, and a continuing series of more
minor leaks and emissions since. Although reprocessing recovers
some of the most intensely radioactive material, it also produces
large amounts of low-level waste, particularly cooling water and
gases, which are released. Discharges from Sellafield have now
made the Irish Sea the world's most radioactive ocean.

erating electricity by nuclear energy. Disposal, by burial on land or at sea, has raised problems of safety, environmental pollution, and security. In absolute terms, nuclear waste cannot be safely relocated or disposed of.

Nuclear waste may have an active life of several thousand years. Reactor waste is of three types: *high-level* spent fuel, or the residue when nuclear fuel has been removed from a reactor and reprocessed; *intermediate*, which may be long- or short-lived; and *low-level*, but bulky, waste from reactors, which has only short-lived radioactivity.

The dumping of nuclear waste at sea officially ceased 1983, when a moratorium was agreed by the members of the London Dumping Convention (a United Nations body that controls disposal of wastes at sea). However, the USSR continued dumping covertly and deposited thousands of tonnes of nuclear waste and three faulty reactors in the sea 1964–86.

Waste from a site where uranium is mined or milled may have an active life of several thousand years, and spent (irradiated) fuel is dangerous for tens of thousands of years. Sea disposal has occurred at many sites, for example 450 km/300 mi off Land's End, England, but there is no guarantee of the safety of this method of disposal, even for low-activity waste. Beneath the sea the containers would start to corrode after 1,000 years, and the cylinders themselves would dissolve within the next 1,000 years. There have been proposals to dispose of high-activity waste in old mines, granite formations, and specially constructed bunkers. One solution favoured by the industry is *vitrification* in which waste is combined with glass to form solid cylinders, which are subsequently stored down deep shafts. However, given enough time, these containers would also start to corrode.

About one-third of the fuel from nuclear reactors becomes spent each year. It is removed to a *reprocessing* plant where radioactive waste products are chemically separated from remaining uranium and plutonium, in an expensive and dangerous process. This practice increases the volume of radioactive waste more than a hundred times.

O

Ocean Drilling Program (ODP, formerly the *Deep-Sea Drilling Project* 1968–85) research project initiated by the USA 1968 to sample the rocks of the ocean crust. The operation became international 1975, when Britain, France, West Germany, Japan, and the USSR also became involved.

Boreholes were drilled in all the oceans using the ships *Glomar Challenger* and JOIDES *Resolution*, and knowledge of the nature and history of the ocean basins was increased dramatically. The technical difficulty of drilling the seabed to a depth of 2,000 m/6,500 ft was overcome by keeping the ship in position with side-thrusting propellers and satellite navigation, and by guiding the drill using a radio location system. The project is intended to continue until 2005.

oceanography study of the oceans. Its subdivisions deal with each ocean's extent and depth, the water's evolution and composition, its physics and chemistry, the bottom topography, currents and wind effects, tidal ranges, the biology, and the various aspects of human use.

Oceanography involves the study of water movements – currents, waves, and tides – and the chemical and physical properties of the seawater. It deals with the origin and topography of the ocean floor – ocean trenches and ridges formed by plate tectonics, and continental shelves from the submerged portions of the continents. Much oceanography uses computer simulations to plot the possible movements of the waters, and many studies are carried out by remote sensing.

The World Ocean Circulation Experiment, a seven-year project begun 1990, involves researchers from 44 countries examining the physics of the ocean and its role in the climate of the Earth. It is based at Southampton University, England.

odour irritating smells, which cause a nuisance but are not actually dangerous. They are a frequent cause of the complaints made by house-holders living near some types of industry or agriculture. Of the 13,500 complaints made to Environmental Health Officers in England and Wales about bad smells, three quarters related to odours from industry, and a quarter to odours from farms. Most of the odours coming from farms are caused by the spreading of sewage slurry.

oil natural mineral oil, a thick greenish-brown flammable liquid found underground in permeable rocks. Crude oil is known as ⇨petroleum.

oil refinery industrial complex where crude oil is processed into different products. The light volatile parts of the oil form ⇨petroleum, while the heavier parts make bitumen and petrochemicals. Oil refineries are often located at deep-water ports or near their industrial markets. They need flat land and water for cooling.

oil spill oil released by damage to or discharge from a tanker or oil installation. An oil spill kills all shore life, clogging up the feathers of birds and suffocating other creatures. At sea toxic chemicals leach into the water below, poisoning sea life. Mixed with dust, the oil forms globules that sink to the seabed, poisoning sea life there as well. Oil spills are broken up by the use of detergents but such chemicals can themselves damage wildlife. The annual spillage of oil is 8 million barrels a year. At any given time tankers are carrying 500 million barrels.

In March 1989 the *Exxon Valdez* (belonging to the ⇨Exxon Corporation) spilled oil in Alaska's Prince William Sound, covering 12,400 sq km/4,800 sq mi and killing at least 34,400 sea birds, 10,000 sea otters, and up to 16 whales. The incident led to the US Oil Pollution Act of 1990, which requires tankers operating in US waters to have double hulls. The world's largest oil spill was in the Persian Gulf in Feb 1991 as a direct result of hostilities during the Gulf War. Around 6–8 million barrels of oil were spilled, polluting 675 km/420 mi of Saudi coastline. In some places, the oil was 30 cm/12 in deep in the sand. The amount of oil entering oceans from shipping operations decreased by 60% 1981–91.

okapi ruminant *Okapia johnstoni* of the giraffe family, although with much shorter legs and neck, found in the tropical rainforests of central

THE *BRAER* OIL SPILL

On 5 Jan 1993, the oil tanker *Braer* ran ashore in heavy seas on the coast of the Scottish Shetland Islands. While salvage experts watched helplessly from the cliffs, several days of intense storms broke the ship completely in two and spilled virtually the whole of its 85,000-tonne cargo into the surrounding sea. Huge waves and gale-force winds dispersed the oil, blowing volatile chemicals over much of the island. Hundreds of islanders experienced headaches, diarrhoea, and eczema, and 1,500 sea birds were washed ashore dead, although the total death toll may have been far higher.

An exclusion zone of 1,000 sq km/400 sq mi was created, fishing within it forbidden, and the fish from salmon farms banned from sale. Salmon that would have been worth £7 million if uncontaminated was been minced and used to feed captive mink in Norway. On shore, areas of sheep pasture are judged off limits because of contamination by storm-blown oil droplets.

The Shetland islanders now face years of legal wrangling about compensation. For many, the issue has as much to do with politics as accidents. Physically and culturally remote from the decision-making power of the government in London, the islanders have had to fight against the risks to fisheries from the nuclear power plant on the mainland. Fears of a major oil-tanker accident had been expressed for years. Since the *Braer* accident, tankers have continued to sail within 16 km/10 mi of the islands. Isobel Mitchell, who organized a petition demanding a full public inquiry into the disaster, signed by a third of the Shetland population, said some months afterwards: 'Nothing has really changed to stop the same thing happening tomorrow.'

Africa. Purplish brown with a creamy face and black and white stripes on the legs and hindquarters, it is excellently camouflaged. Okapis have remained virtually unchanged for millions of years. Only a few hundred are now thought to survive.

Oman *the country has extensively modernized over recent years and is a dynamic fast expanding society, which has put further pressure on its already inadequate water supply. Recurrent droughts and inadequate rainfall in agricultural areas have led to a heavy dependence on* *irrigation from underground ⇨aquifers. However, these are shallow and in danger of being exhausted. In the north, sea water has contaminated the aquifers as a result of overexploitation. Oil pollution from discharges in the Gulf has contaminated many beaches, threatening marine and bird life.*

opencast mining or *open-pit mining* or *strip mining* mining at the surface rather than underground. Coal, iron ore, and phosphates are often extracted by opencast mining. Often the mineral deposit is covered by soil, which must first be stripped off, usually by large machines such as walking draglines and bucket-wheel excavators. The ore deposit is then broken up by explosives.

One of the largest excavations in the world has been made by opencast mining at the Bingham Canyon copper mine in Utah, USA, measuring 790 m/2,590 ft deep and 3.7 km/2.3 mi across.

orang-utan ape *Pongo pygmaeus*, found solely in Borneo and Sumatra. Up to 1.65 m/5.5 ft in height, it is covered with long, red-brown hair and lives mainly a solitary, arboreal life, feeding chiefly on fruit. Now an endangered species, it is officially protected because its habitat is being systematically destroyed by ⇨deforestation. Orangutans are slow-moving and have been hunted for food, as well as by animal collectors.

organic compound in biochemistry, one of the class of compounds whose behaviour is influenced by the chemistry of carbon. All organic compounds contain carbon, which was once thought to be present only in living things. The original distinction between organic and inorganic compounds was based on the belief that the molecules of living systems were unique, and could not be synthesized in the laboratory. Today it is routine to manufacture thousands of organic chemicals both in research and in the drug industry.

organic farming farming without the use of synthetic fertilizers (such as ⇨nitrates and phosphates) or ⇨pesticides (herbicides, insecticides, and fungicides) or other agrochemicals (such as hormones, growth stimulants, or fruit regulators).

In place of artificial fertilizers, compost, manure, seaweed, or other substances derived from living things are used (hence the name 'organic'). Growing a crop of a nitrogen-fixing plant such as lucerne, then ploughing it back into the soil, also fertilizes the ground. Some organic farmers use naturally occurring chemicals such as nicotine or pyrethrum to kill pests, but control by nonchemical methods is preferred. Those methods include removal by hand, intercropping (planting with companion plants which deter pests), mechanical barriers to infestation, crop rotation, better cultivation methods, and biological control. Weeds can be controlled by hoeing, mulching (covering with manure, straw, or black plastic), or burning off. Organic farming methods produce food with minimal pesticide residues and greatly reduce pollution of the environment. They are more labour intensive, and therefore more expensive, but use less fossil fuel. Soil structure is greatly improved by organic methods, and recent studies show that a conventional farm can lose four times as much soil through erosion as an organic farm, although the loss may not be immediately obvious.

ortolan songbird *Emberiza hortulana* of the bunting family, common in Europe and W Asia, migrating to Africa in the winter. Long considered a delicacy among gourmets, it has become rare and is now a protected species.

oryx any of the genus *Oryx* of large antelopes native to Africa and Asia. The Arabian oryx *O.leucoryx*, at one time extinct in the wild, has been successfully reintroduced into its natural habitat using stocks bred in captivity.
scimitar oryx Medium-sized oryx *Oryx dammah*. Its startling, swept-back horns have made it a prime target for hunters. It was once found over virtually the whole of the Sahel. Numbers started to decline sharply in the 1950s and the animals were reduced to scattered groups by the 1970s, then reached the edge of extinction by the 1980s. The initial cause of decline was the destruction of grasslands, but savage hunting annihilated the remainder. There are captive breeding herds and plans to reintroduce the species in Tunisia.

otter any of various aquatic carnivores of the weasel family, found on all continents except Australia. Otters have thick, brown fur, short limbs, webbed toes, and long, compressed tails. They are social, playful, and agile.

In the UK, otters have been driven to the brink of extinction by hunting and, more importantly, by the pollution and destruction of their riverside habitat as a result of agricultural and commercial activity. They are now slowly making a recovery with the aid of protective legislation and are being reintroduced to many river systems by naturalists and conservation organizations, although the population is still fragile in many areas of the country.

overfishing fishing at rates that exceed the ⇨sustained-yield cropping of fish species, resulting in a net population decline. For example, in the North Atlantic, herring has been fished to the verge of extinction and the cod and haddock populations are severely depleted. In the Third World, use of huge factory ships, often by fisheries from industrialized countries, has depleted stocks for local people who cannot obtain protein in any other way.

Ecologists have long been concerned at the wider implications of overfishing, in particular the devastation wrought on oceanic ⇨food chains. Massive sea-bird wrecks, in which dead sea birds are washed up on the coasts, their cadavers revealing signs of starvation, indicate a system in crisis. For example, in January 1994, tens of thousands of dead sea birds were washed up on the UK coastline of the North Sea. They are believed to have died of starvation as a result of overfishing. Estimates that the fishing catch could in principle be increased, perhaps to 100 million tonnes per annum, depend on better management of fishing programmes; it is estimated that, annually, 20 million tonnes of fish are discarded from fishing vessels at sea, because they are not the species sought.

The United Nations Food and Agriculture Organization estimates that worldwide overfishing has damaged oceanic ecosystems to such an extent that potential catches are on average reduced by 20%. The global catch peaked at 86 million tonnes 1989 but had fallen to 80 million tonnes 1992. This is in spite of the use of more powerful trawlers and bigger nets. Some species are particularly badly effected, for example, the catch of the Atlantic cod, the Cape hake, the haddock and the silver hake has fallen from 5 million tonnes in 1970 to 2.6 million tonnes in 1989. The cod is protected by partial fishing moratoria off Canada and in European waters, but stocks may take years to recover.

overpopulation too many people for the resources available in an area (such as food, land, and water). The consequences were first set out in the ⇨Malthus theory.

Although there is often a link between overpopulation and population density, high densities will not always result in overpopulation. In many countries, resources are plentiful and the infrastructure and technology are well developed. This means that a large number of people can be supported by a small area of land. In some developing countries, such as Bangladesh, Ethiopia, and Brazil, insufficient food, minerals, and energy, and inequitable income distribution result in poverty and often migration in search of better living conditions. Here even low population densities may amount to overpopulation. Overpopulation may also result from a decrease in resources or an increase in population or a combination of both.

oxygen colourless, odourless, tasteless, nonmetallic, gaseous element, symbol O, atomic number 8, relative atomic mass 15.9994. It is the most abundant element in the Earth's crust (almost 50% by mass), forms about 21% by volume of the atmosphere, and is present in combined form in water and many other substances.

Earth contained no oxygen when it formed some 4.5 billion years ago. All free oxygen found in today's atmosphere originates from photosynthetic organisms. Most organisms alive today are ⇨aerobic organisms – that is, they use oxygen for oxidizing glucose in the energy-giving process of ⇨respiration.

Oxygen is very reactive and combines with all other elements except the inert gases and fluorine. It is present in carbon dioxide, silicon dioxide (quartz), iron ore, calcium carbonate (limestone). In nature it exists as a molecule composed of two atoms (O_2); single atoms of oxygen are very short-lived owing to their reactivity. They can be produced in electric sparks and by the Sun's ultraviolet radiation in space, where they rapidly combine with molecular oxygen to form ⇨ozone (an allotrope of oxygen).

ozone O_3 highly reactive pale-blue gas with a penetrating odour. Ozone is an allotrope of oxygen, made up of three atoms of oxygen. It is formed when the molecule of the stable form of oxygen (O_2) is split by ultraviolet radiation or electrical discharge.

At ground level, ozone can cause asthma attacks, stunted growth in plants, and corrosion of certain materials. It is produced by the action of sunlight on air pollutants, including car exhaust fumes, and is a major air pollutant in hot summers. In the upper atmosphere ozone has a beneficial effect, shielding life on Earth from ultraviolet rays, a cause of skin cancer. Ozone is a powerful oxidizing agent and is used industrially in bleaching and air conditioning.

A continent-sized hole has formed over Antarctica as a result of damage to the ozone layer. This has been caused in part by ⇨chlorofluorocarbons (CFCs), but many reactions destroy ozone in the stratosphere: nitric oxide, chlorine, and bromine atoms are implicated. In 1989 ozone depletion was 50% over the Antarctic compared with 3% over the Arctic. In April 1991 satellite data from NASA revealed that the ozone layer had depleted by 4–8% in the N hemisphere and by 6–10% in the S hemisphere between 1978 and 1990. It is believed that the ozone layer is depleting at a rate of about 5% every 10 years over N Europe, with depletion extending south to the Mediterranean and southern USA. However, ozone depletion over the polar regions is the most dramatic manifestation of a general global effect. See also ⇨Montréal Protocol.

As a pollutant at ground level, ozone is so dangerous that the US Environment Protection Agency recommends people should not be exposed for more than one hour a day to ozone levels of 120 parts per billion (ppb), while the World Health Organization recommends a lower 76–100 ppb. It is known that even at levels of 60 ppb ozone causes respiratory problems, and may cause the yields of some crops to fall. In the USA the annual economic loss due to ozone has been estimated at $5.4 billion.

ozone depleter any chemical that destroys the ozone in the stratosphere. Most ozone depleters are chemically stable compounds containing chlorine or bromine, which remain unchanged for long enough to drift up to the upper atmosphere.

Once in the upper atmosphere, they are broken up by the intense solar radiation and form a cocktail of more active substances which then

react with ozone, causing its depletion. The best known are
⇨chlorofluorocarbons (CFCs), but many other ozone depleters are
known, including halons, used in some fire extinguishers; methyl chloroform and carbon tetrachloride, both solvents; some CFC substitutes;
and the pesticide methyl bromide. Most research into alternatives to
ozone depleters seeks chemical alternatives which will break up before
they get into the upper atmosphere, but still have a useful working life
as a refrigerant or propellant.

P

packaging material, usually of metal, paper, plastic, or glass, used to protect products, make them easier to display, and as a form of advertising. Environmentalists have criticized packaging materials as being wasteful of energy and resources. Recycling bins are being placed in residential areas to facilitate the collection of surplus packaging.

In Germany from Dec 1991, shoppers have been able to leave behind surplus paper and plastic packaging. From 1993, shops have been obliged to take back empty cans, toothpaste tubes, and yogurt cartons, and a compulsory deposit is levied on certain types of packaging.

In the UK, over 5 billion cans of drink are sold, although only 2% of metal cans are recycled.

panda one of two carnivores of different families, native to NW China and Tibet. The *giant panda* Ailuropoda melanoleuca has black-and-white fur with black eye patches and feeds mainly on bamboo shoots, consuming about 8 kg/17.5 lb of bamboo per day. It can grow up to 1.5 m/4.5 ft long, and weigh up to 140 kg/300 lb. The *lesser* or *red panda* Ailurus fulgens, of the raccoon family, is about 50 cm/1.5 ft long, and is black and chestnut, with a long tail.

Giant pandas are an endangered species, partly due to their reliance on one source of food, bamboo, which is becoming rarer in some areas as a result of encroaching urbanization. Destruction of the giant pandas' natural habitats threatens to make them extinct in the wild, and they are the focus of many conservation efforts.

paper thin, flexible material made in sheets from vegetable fibres (such as wood pulp) or rags and used for writing, drawing, printing, packaging, and various household needs. Today most paper is made from wood pulp on a Foudrinier machine, then cut to size. Paper products absorb 35% of the world's annual commercial wood harvest; recycling avoids some of the enormous waste of trees, and most paper-makers plant and replant their own forests of fast-growing stock.

Papua New Guinea *the country's rich biodiversity is under threat from illegal logging, which is being carried out at an unsustainable rate, is destroying large areas of wildlife habitat, and displacing indigenous populations. ⇨Open-cast mining has also caused severe environmental damage, particularly contamination of surrounding land. In 1990, a copper mine in Bougainvillaea was closed after outcry from local people at the damage it was causing, although the area is still heavily polluted from other mines still operating.*

paraquat $CH_3(C_5H_4N)_2CH_{3.2}CH_3SO_4$ (technical name *1,1-dimethyl-4,4-dipyridylium*) nonselective herbicide (weedkiller). Although quickly degraded by soil microorganisms, it is deadly to human beings if ingested.

park and ride town-planning scheme in which parking space is provided (often free) some distance away from the central business district. Shoppers are taken by bus to the central area, which may be traffic-free (⇨pedestrianization). Park and ride is one of the planning strategies that can be used to combat congestion. The scheme became widespread in the UK during the 1980s.

parrot any bird of the order Psittaciformes, abundant in the tropics, especially in Australia and South America. They are mainly vegetarian, and range in size from the 8.5 cm/3.5 in pygmy parrot to the 100 cm/40 in Amazon macaw.

Several species are endangered. One of the rarest is the imperial parrot, found only in Dominica in the Caribbean, which is threatened by deforestation. Parrots are also under threat from the pet trade as they have been valued as pets in the Western world for many centuries. Every year some 600,000 parrots are caught in the wild for sale but many die in transit.

peat fibrous organic substance found in bogs and formed by the incomplete decomposition of plants such as sphagnum moss. N Asia, Canada, Finland, Ireland have large deposits, which have been dried and used as fuel from ancient times. Peat can also be used as a soil additive.

Peat bogs began to be formed when glaciers retreated, about 9,000 years ago. They grow at the rate of only a millimetre a year, and large-scale digging can result in destruction both of the bog and of specialized plants growing there.

In 1990 the third largest peat bog in Britain, on the borders of Shropshire and Clwyd, was bought by the Nature Conservancy Council, the largest purchase ever made by the NCC.

pedestrianization the closing of an area to traffic, making it more suitable for people on foot. It is now common in many town shopping centres, since cars and people often obstruct one another. This restricts accessibility and causes congestion. Sometimes service vehicles (such as buses and taxis) are allowed access.

Peru *an estimated 38% of the 8,000 sq km/3,100 sq mi of coastal lands under irrigation are either waterlogged or suffering from saline water. Only half the population has access to clean drinking water. ⇨Overfishing and discharge of industrial and domestic waste pose a threat to Peru's anchovy fishery.*

---

pesticide any chemical used in farming, gardening or indoors to combat pests. Pesticides are of three main types: ***insecticides*** (to kill insects), ***fungicides*** (to kill fungal diseases), and ***herbicides*** (to kill plants, mainly those considered weeds). Pesticides cause a number of pollution problems through spray drift onto surrounding areas, direct contamination of users or the public, and as residues on food.

The safest pesticides include those made from plants, such as the insecticides pyrethrum and derris. More potent are synthetic products, such as chlorinated hydrocarbons. These products, including ⇨DDT and dieldrin, are highly toxic to wildlife and often to human beings, so their use is now banned or restricted by law in some areas and is declining. Safer pesticides, such as malathion, are based on organic phosphorus compounds, but they still present hazards to health. The aid organization Oxfam estimates that pesticides cause about 10,000 deaths worldwide every year.

Pesticides were used to deforest SE Asia during the Vietnam War, causing death and destruction to the area's ecology and lasting health and agricultural problems. In 1991 Central America was the world's highest consumer of pesticides per head of the population. There are around 4,000 cases of acute pesticide poisoning a year in the UK. The global trade in pesticides and related products is estimated to be worth $27 billion per annum, with sales to Asia and Latin America forming a massive trade flow.

petrochemical chemical derived from the processing of petroleum (crude oil).

Petrochemical industries are those that obtain their raw materials from the processing of petroleum and natural gas. Polymers, detergents, solvents, and nitrogen fertilizers are all major products of the petrochemical industries. Inorganic chemical products include carbon black, sulphur, ammonia, and hydrogen peroxide.

In the UK, ICI has a large petrochemical plant on Teesside.

petrol engine the most commonly used source of power for motor vehicles, introduced by the German engineers Gottlieb Daimler and Karl Benz 1885. The petrol engine is a complex piece of machinery made up of about 150 moving parts. It is a reciprocating piston engine, in which a number of pistons move up and down in cylinders. The motion of the pistons rotate a crankshaft, at the end of which is a heavy flywheel. From the flywheel the power is transferred to the car's driving wheels via the transmission system of clutch, gearbox, and final drive.

The particular conditions of combustion in such an engine produce copious amounts of the polluting gases carbon dioxide and the nitrogen dioxides. In the *lean-burn* engine the ratio of petrol to air is reduced, with adaptations made to the cylinder structure to aid the combustion of such a weak mixture. This gives greater fuel efficiency and, because of the lower temperature of combustion, reduced emission of the polluting nitrogen oxides. However, these engines are not as yet in commercial production.

The parts of the petrol engine can be subdivided into a number of systems. The *fuel system* pumps fuel from the petrol tank into the carburettor. There it mixes with air and is sucked into the engine cylinders. (With electronic fuel injection, it goes directly from the tank into the cylinders by way of an electronic monitor.) The *ignition system* supplies the sparks to ignite the fuel mixture in the cylinders. By means of an ignition coil and contact breaker, it boosts the 12-volt battery voltage to pulses of 18,000 volts or more. These go via a distributor to the spark plugs in the cylinders, where they create the sparks. (Electronic ignitions replace these parts.) Ignition of the fuel in the cylinders produces temperatures of 700°C/1,300°F or more, and the engine must be cooled to prevent overheating.

Most engines have a *water-cooling system*, in which water circulates through channels in the cylinder block, thus extracting the heat. It flows through pipes in a radiator, which are cooled by fan-blown air. A few cars and most motorbikes are air-cooled, the cylinders being surrounded by many fins to present a large surface area to the air. The *lubrication system* also reduces some heat, but its main job is to keep the moving parts coated with oil, which is pumped under pressure to the camshaft, crankshaft, and valve-operating gear.

petroleum or *crude oil* natural mineral oil, a thick greenish-brown flammable liquid found underground in permeable rocks. Petroleum consists of hydrocarbons mixed with oxygen, sulphur, nitrogen, and other elements in varying proportions. It is thought to be derived from ancient organic material that has been converted by, first, bacterial action, then heat and pressure (but its origin may be chemical also). From crude petroleum, various products are made by distillation and other processes; for example, fuel oil, petrol, kerosene, diesel, lubricating oil, paraffin wax, and petroleum jelly.

The USA led in production until the 1960s, when the Middle East outproduced other areas, their immense reserves leading to a worldwide dependence on cheap oil for transport and industry. In 1961 the Organization of the Petroleum Exporting Countries (OPEC) was established to avoid exploitation of member countries; after OPEC's price rises in 1973, the International Energy Agency (IEA) was established 1974 to protect the interests of oil-consuming countries. New technologies were introduced to pump oil from offshore and from the Arctic (the Alaska pipeline) in an effort to avoid a monopoly by OPEC. Global consumption of petroleum in 1993 was 23 billion barrels.

Petroleum products and chemicals are used in large quantities in the

manufacture of detergents, artificial fibres, plastics, insecticides, fertilizers, pharmaceuticals, toiletries, and synthetic rubber. Aviation fuel is
a volatile form of petrol.

petroleum: chronology

1859	Edwin Drake drilled the world's first successful oil well in Titusville, Pennsylvania, USA, to a depth of 18 m/70 ft.
1865	The first oil pipeline, 9,750 m/32,000 ft long, was constructed at Oil Creek, Pennsylvania, USA, to carry oil from well to nearby coalfield.
1896	The first offshore wells were drilled from piers off the California coast.
1899	The first gravity meter was produced.
1914	Reginald Fessenden patented the seismograph.
1939	The aircraft-borne magnetometer was developed to measure magnetism of rocks.
1966	Oil was discovered beneath the North Sea.
1967	The worst oil spill in British waters from the *Torrey Canyon*, which struck rocks at Lands End. Over 108,000 tonnes cf oil were spilled.
1974	The world's deepest oil well, 10,941 m/31,441 ft, was drilled in Oklahoma, USA.
1979	The oil rig Ixtoc I in the Gulf of Mexico accidentally released 545,000 tonnes of oil into the sea. The slick spread for 640 km/400 mi. Later the same year, the worst tanker spillage occurred. Two tankers, the *Atlantic Empress* and the *Aegean Captain*, collided off the island of Tobago, in the Caribbean Sea. More than 230,000 tonnes of oil were spilled.
1984	An exploratory well was drilled off the coast of New England, USA, in water of depth 2,116 m/6,942 ft – a world record.
1988	The Piper Alpha drilling rig in the North Sea caught fire in July, killing 167 people.
1989	The worst spill in American waters occurred when 55,000 tonnes of oil escaped from the *Exxon Valdez* off the Alaskan coast, near Prince William Sound.
1991	The worst spill to date was a consequence of the Gulf War when Iraqi forces opened the pipeline into the Persian Gulf and coalition forces caused further damage during Operation Desert Storm.

The burning of fuels derived from petroleum is a major cause of air
pollution. Its transport can lead to major environmental catastrophes –
for example, the *Torrey Canyon* tanker lost off SW England, 1967,

which led to an agreement by the international oil companies 1968 to pay compensation for massive shore pollution. The 1989 ⇨oil spill in Alaska from the *Exxon Valdez* damaged the area's fragile environment, despite clean-up efforts. Drilling for petroleum involves the risks of accidental spillage and drilling-rig accidents. The problems associated with petroleum have led to the various ⇨alternative energy technologies.

A new kind of bacterium was developed during the 1970s in the USA, capable of 'eating' oil as a means of countering oil spills. Its creation gave rise to the so-called Frankenstein law (a ruling that allowed new forms of life created in laboratories to be patented).

pH scale from 0 to 14 for measuring acidity or alkalinity. A pH of 7 indicates neutrality, below 7 is acid, while above 7 is alkaline. Strong acids, such as those used in car batteries, have a pH of about 2; strong alkalis such as sodium hydroxide are pH 13.

The pH of a solution can be measured by using a broad-range indicator, either in solution or as a paper strip. The colour produced by the indicator is compared with a colour code related to the pH value. An alternative method is to use a pH meter fitted with a glass electrode.

Acidic fruits such as citrus fruits are about pH 4. Fertile soils have a pH of about 6.5 to 7.0, while weak alkalis such as soap are 9 to 10. ⇨Acid rain typically has a pH of 4–4.7, although the lowest recorded in the UK was pH 2.4 in Pitlochry, Scotland. However, it is estimated that the London ⇨smog 1952, which killed 4,000 people, had a pH value of 1.7, as strong as battery acid.

Philippines *cleared for timber, tannin, and the creation of fish ponds, the mangrove forest was reduced from 5,000 sq km/1,930 sq mi to 380 sq km/146 sq mi between 1920 and 1988. Illegal logging remains a serious threat to biodiversity and wildlife habitat. Pollution is also a problem, with air pollution in urban areas being so severe as to be a health hazard and nearly 40 river systems across the country polluted by toxic waste.*

---

phosphorus highly reactive, nonmetallic element, symbol P, atomic number 15, relative atomic mass 30.9738. It occurs in nature as phosphates (commonly in the form of the mineral apatite), and is essential to plant and animal life. Compounds of phosphorus are used in commercial fertilizers, various organic chemicals, for matches and fireworks, and in glass and steel.

The element has three allotropic forms: a black powder; a white-yellow, waxy solid that ignites spontaneously in air to form the poisonous gas phosphorus pentoxide; and a red-brown powder that neither ignites spontaneously nor is poisonous.

photosynthesis process by which green plants trap light energy and use it to drive a series of chemical reactions, leading to the formation of carbohydrates. All animals ultimately depend on photosynthesis because it is the method by which the basic food (sugar) is created. Photosynthesis by cyanobacteria (⇨blue-green algae) was responsible for the appearance of oxygen in the Earth's atmosphere 2 billion years ago, and photosynthesis by plants maintains the oxygen level today.

For photosynthesis to occur, the plant must possess chlorophyll and must have a supply of carbon dioxide and water. Actively photosynthesizing green plants store excess sugar as starch.

photovoltaics solid state, silicon devices which convert sunlight directly into energy. They need have no moving parts, operate quietly and without emissions, have a long life-time and require little or no maintenance. They are popularly known as solar cells and are encountered most frequently in sun-powered pocket calculators, camera light metres, and a range of high-cost gadgets.

pioneer species those species that are the first to colonize and thrive in new areas. Coal tips, recently cleared woodland, and new roadsides are areas where pioneer species will quickly appear. As the habitat matures other species take over, a process known as *succession*.

Piper Alpha disaster accident aboard the North Sea oil platform Piper Alpha on 6 July 1988, in which 167 people died. The rig was devastated by a series of explosions, caused initially by a gas leakage. An official inquiry held into the disaster highlighted the vulnerability of offshore rigs.

plastic any of the stable synthetic materials that are fluid at some stage in their manufacture, when they can be shaped, and that later set to rigid or semi-rigid solids. Plastics today are chiefly derived from petroleum. Most are polymers, made up of long chains of identical molecules.

Processed by extrusion, injection-moulding, vacuum-forming and compression, plastics emerge in consistencies ranging from hard and inflexible to soft and rubbery. They replace an increasing number of natural substances, being lightweight, easy to clean, durable, and capable of being rendered very strong – for example, by the addition of carbon fibres – for building aircraft and other engineering projects.

Biodegradable plastics are increasingly in demand: Biopol was developed in 1990. Soil microorganisms are used to build the plastic in their cells from carbon dioxide and water (it constitutes 80% of their cell tissue). The unused parts of the microorganism are dissolved away by heating in water. The discarded plastic can be placed in landfill sites where it breaks back down into carbon dioxide and water. It costs three to five times as much as ordinary plastics to produce. Another plastic digested by soil microorganisms is polyhydroxybutyrate (PHB), which is made from sugar.

plutonium silvery-white, radioactive, metallic element of the actinide series, symbol Pu, atomic number 94, relative atomic mass 239.13. It occurs in nature in minute quantities in pitchblende and other ores, but is produced in quantity only synthetically. It has six allotropic forms and is one of three fissile elements (elements capable of splitting into other elements – the others are thorium and uranium). The element has awkward physical properties and is the most toxic substance known.

Because Pu-239 is so easily synthesized from abundant uranium, it has been produced in large quantities by the weapons industry. It has a long ⇨half-life (24,000 years) during which time it remains highly toxic. Plutonium is dangerous to handle, difficult to store, and impossible to dispose of.

Poland *atmospheric pollution derived from coal (producing 90% of the country's electricity), toxic waste from industry, and lack of sewage treatment have resulted in the designation of 27 ecologically endangered areas. Half the country's lakes have been seriously contaminated and three-quarters of its drinking water does not meet official health standards.*

pollarding type of pruning whereby the young branches of a tree are severely cut back, about 2–4 m/6–12 ft above the ground, to produce a stump-like trunk with a rounded, bushy head of thin new branches. It is similar to ⇨coppicing. It is often practised on willows, where the new branches or 'poles' are used for fencing and firewood. Pollarding is also used to restrict the height of many street trees.

polluter-pays principle the idea that whoever causes pollution is

pollution: chronology

1709	English industrialist Abraham Darby first smelted iron ore with coke, massively increasing industry's use of fossil fuels thereafter.
1852	Scottish chemist Robert Angus Smith first recorded unusually acidic rainfall in Manchester.
1856	The 'Year of the Great Stink' in London; the River Thames smelled so badly that parliament was evacuated.
1952	London afflicted by severe smogs ('pea-soupers') killing 4,000.
1962	US natural scientist Rachel Carson published Silent Spring, a treatise against the ecological dangers of pesticides.
1971	First international pollution compensation fund, the International Maritime Organization Oil Pollution Fund, established.
1984	A leak from a US-owned pesticide plant in Bhopal, India, killed 3,000, with another 30,000 seriously injured.
1985	A 'hole' in the stratospheric ozone layer over the Antarctic first reported.
1986	Accident in the nuclear power plant at Chernobyl, Ukraine, contaminates large area of USSR. Fire at a chemical plant in Basel, Switzerland, released chemicals into the River Rhine, killing almost all living organisms in a 320 km stretch between Basel and Mainz.
1989	The tanker *Exxon Valdez* was wrecked off Alaska and discharged 37,000 tonnes of crude oil, causing substantial damage to bird and aquatic life and the Alaskan coast.
1991	Iraqi military discharged 7–8 million barrels of oil in Kuwait, nearly twice the size of the previous largest oil spill.
1992	UN Convention on Climate Change opened for signing.
1993	Wrecked tanker *Braer* discharged 85,000 tonnes of oil on the Scottish coast.
1994	First international deal in which states agreed to reduce pollutants by differing degrees according to the amount of damage they cause was signed.

POLLUTION IN EASTERN EUROPE

One of the main factors contributing to the collapse of Soviet-style communism in E Europe was growing public discontent about the state of the environment. The Polish Ecology Club was the first legally independent organization to be established in Poland after the rise of Solidarity. Environmental issues were rallying calls to the former opposition in Czechoslovakia and Hungary, and Hungarian opposition to a dam on the Danube drew international attention. Yet now that the countries of central and E Europe are opening up for the first time in decades, the overall state of the environment is by no means clear, as was once assumed by people in the West. E Europe may have hug environmental problems but the news is not all bad.

Problems

The best known environmental problems in the central European countries are connected with various forms of pollution. Inefficient industrial processes, lax environmental standards, and the intrinsically higher pollution levels from the sulphur-rich lignite coal found in some of the former communist countries have combined to cause extremely high air-pollution levels in places. For example, in the mid-1980s Poland had annual emissions of sulphur and nitrogen oxides almost identical to those coming from the UK but with a population not much more than half the size of Britain's, meaning that there was roughly twice as much acidic pollution per person. East Germany emitted more air pollution than West Germany, despite having not much more than a quarter of the population. Yet Britain and W Germany had at that time some of the highest air-pollution emissions of W European countries.

Signs of damage

The results for the East include severely damaged forests, crumbling historical buildings, as in the heavily polluted Polish city of Kraków, and evidence that human cancer levels have increased in areas of high pollution. Children are kept home from school on 'bad days' in the most polluted areas of Czechoslovakia.

There are other serious pollution problems. Freshwater pollution is reaching crisis levels in some areas. The Volga–Caspian basin, source of most of the world's caviar, contains 700 times the legal levels of petroleum products and 113 times the permitted levels of surface pollution from synthetic products. This has led to fears of 'ecological disaster', according to Professor Vladimir Lukyanenko, a top Russian biologist. Large areas of the Siberian taiga are said to be polluted by oil from spillage during drilling. Children in the Bulgarian town of Kuklen have blood lead levels so high that they would be sent for detoxification in the USA. Infant mortality rates in Czechoslovakia are 60% higher than in the West.

Signs of hope

Yet in other ways ecological conditions in E Europe are superior to those found in the West. The largest remaining areas of natural or ancient forest in Europe are found in the East, where they serve as a vital reservoir of biodiversity. There are large populations of rare animals, such as bear and wolf.

Unfortunately, these irreplaceable forests are now being slated for logging, often by western companies moving into rich new markets in the East. Russia has some 40% of the world's temperate timber, mainly in Siberia, where plans to log huge areas of this are rapidly gaining pace. Latvia is already developing forestry in areas that would be classed as internationally important nature reserves in most other European countries.

Help or hindrance from the West

The West is no mere spectator to the problems of E Europe; it is playing a role in the increasing damage from agrochemical pollution. Although some parts of E Europe have long been run with extremely intensive agricultural systems, other areas are effectively still practising peasant farming. These places have a chance to develop organic and low-input farming systems with produce that would find a ready market in the West. However, Western aid funding and the agrochemical companies are pushing the development of the type of intensive agriculture that is gradually being abandoned, or at least severely modified, in the rest of Europe.

E Europe is an ecological disaster area in some respects. However, in other ways it has the potential to provide a reservoir of biodiversity that has long been destroyed in most W European countries. It is important that the 'help' coming from the West does not destroy the good in the process of trying to correct the wrong.

responsible for the cost of repairing any damage. The principle is accepted in British law but has in practice often been ignored; for example, farmers causing the death of fish through slurry pollution have not been fined the full costs of restocking the river.

pollution the harmful effect on the environment of by-products of human activity, principally industrial and agricultural processes – for example, noise, smoke, car emissions, chemical and radioactive effluents in air, seas, and rivers, pesticides, radiation, sewage (see ⇨sewage disposal), household waste, and even visual pollution (principally referring to ugly buildings). Pollution contributes to ozone depletion and the ⇨greenhouse effect.

Pollution control involves higher production costs for the industries concerned, but failure to implement adequate controls may result in irreversible environmental damage and an increase in the incidence of diseases such as cancer. Radioactive pollution results from inadequate ⇨nuclear safety. Polluting gases such as sulphur dioxide can travel on prevailing wind for up to 500 km/300 mi a day. This means that pollution generated in one country can easily affect another and hence it is known as transboundary pollution. Natural disasters may also cause pollution; volcanic eruptions, for example, cause ash to be ejected into the atmosphere and deposited on land surfaces.

In the UK in 1987 ⇨air pollution caused by carbon monoxide emission from road transport was measured at 5.26 million tonnes. In Feb 1990 the UK had failed to apply 21 European Community laws on air and water pollution and faced prosecution before the European Court of Justice on 31 of the 160 EC directives in force.

polychlorinated biphenyl (PCB) any of a group of chlorinated isomers of biphenyl (C_6H_5)$_2$. They are dangerous industrial chemicals, valuable for their fire-resisting qualities. They constitute an environmental hazard because of their persistent toxicity. Since 1973 their use has been limited by international agreement.

polyethylene or *polythene* polymer of the gas ethylene (technically called ethene, C_2H_4). It is a tough, white, translucent, waxy thermoplastic (which means it can be repeatedly softened by heating). It is used for packaging, bottles, toys, wood preservation, electric cable, and pipes.

polystyrene type of ⇨plastic used in kitchen utensils or, in an expanded form, in insulation and ceiling tiles. CFCs are used to produce expanded polythene but alternatives are now being sought.

Polythene trade name for a variety of ⇨polyethylene.

population in biology and ecology, a group of animals of one species, living in a certain area and able to interbreed; the members of a given species in a ⇨community of living things.

population the number of people inhabiting a country, region, area, or town. Population statistics are derived from many sources; for example, through the registration of births and deaths, and from censuses of the population. The first national censuses were taken in 1800 and 1801 and provided population statistics for Italy, Spain, the UK, Ireland, and the USA; and the cities of London, Paris, Vienna, Berlin, and New York.

Since that time a growing number of countries have taken regular censuses, often at ten-yearly intervals, including Austria (1821), France (1821), China (1851), Russia (1861), Japan (1871), and India (1901). Although censuses are approximately accurate for wealthy industrial countries, this may not be the case with other countries. In May 1991, the UN Population Fund (UNFPA) warned that the world's population was growing much faster than predicted, by 250,000 a day, and was expected to reach 5.4 billion in mid-1991, 6.4 billion in 2001, and 10.2 billion in 2060.

population control measures taken by some governments to limit the growth of their countries' populations by trying to reduce ⇨birth rates. Propaganda, freely available contraception, and tax disincentives for large families are some of the measures that have been tried.

The population-control policies introduced by the Chinese government are the best known. In 1979 the government introduced a 'one-child policy' that encouraged ⇨family planning and penalized those couples who have more than one child. The policy has been only partially successful since it has been difficult to administer, especially in rural areas, and has in some cases led to the killing of girls in favour of sons as heirs.

population density the number of people living in a given area, usually expressed as people per square kilometre. It is calculated by dividing the population of a region by its area.

Population density provides a useful means for comparing population distribution. Densities vary considerably throughout the world. High population densities do not always indicate ⇨overpopulation; they do so only where resources are scarce.

population explosion the rapid and dramatic rise in world population that has occurred over the last few hundred years. Between 1959 and 1990, the world's population increased from 2.5 billion to over 5 billion people. It is estimated that it will be at least 6 billion by the end of the century. Most of this growth is now taking place in the ⇨developing world, where rates of natural increase are much higher than in richer countries. Concern that this might lead to ⇨overpopulation has led some countries to adopt ⇨population-control policies.

power station building where electrical energy is generated from a fuel or from another form of energy. Fuels used include fossil fuels such as coal, gas, and oil, and the nuclear fuel uranium. Renewable sources of energy include gravitational potential energy, used to produce ⇨hydroelectric power, and ⇨wind power.

The energy supply is used to turn ⇨turbines either directly by means of water or wind pressure, or indirectly by steam pressure, steam being generated by burning fossil fuels or from the heat released by the fission of uranium nuclei. The turbines in their turn spin alternators, which generate electricity at very high voltage. Conventional coal and oil-fired power stations convert only about 40% of the fuel energy into electrical energy, the rest being wasted as heat. In modern ⇨combined heat and power generation stations fuel efficiency rises to 80% because heat is not wasted but rather is used to warm local industrial estates and housing.

The largest power station in Europe is the Drax power station near Selby, Yorkshire, which supplies 10% of Britain's electricity.

pressurized water reactor (PWR) a ⇨nuclear reactor design used in nuclear power stations in many countries, and in nuclear-powered submarines. In the PWR, water under pressure is the coolant and ⇨moderator. It circulates through a steam generator, where its heat boils water to provide steam to drive power ⇨turbines.

prior informed consent informal policy whereby companies who sell ⇨pesticides to developing countries agree to suspend exporting the product if there is an objection from the government of the receiving country and inform the government of the nature of the pesticide. The situation arises frequently because some pesticides banned in the developed world may be bought by agricultural operations or companies in the Third World, perhaps unaware of any health implications. The policy was adopted by the FAO in 1989, and has since been made binding by the EC on its member states.

propane C_3H_8 gaseous hydrocarbon of the alkane series, found in petroleum and used as fuel.

propene $CH_3CH:CH_2$ (common name ***propylene***) second member of the alkene series of hydrocarbons. A colourless, flammable gas, it is widely used by industry to make organic chemicals, including polypropylene plastics.

proton a positively charged subatomic particle, a constituent of the nucleus of all atoms. It belongs to the baryon subclass of the hadrons. A proton is extremely long-lived, with a lifespan of at least 10^{32} years. It carries a unit positive charge equal to the negative charge of an ⇨electron. Its mass is almost 1,836 times that of an electron, or 1.67×10^{-24} g. The number of protons in the atom of an element is equal to the atomic number of that element.

puma also called ***cougar*** or ***mountain lion*** large wild cat *Felis concolor* found in North and South America. It is tawny coated and is 1.5 m/ 4.5 ft long with a 1 m/3 ft tail. Cougars live alone, with each male occupying a distinct territory; they eat deer, rodents, and cattle. They have been hunted nearly to extinction.

pumped storage hydroelectric plant that uses surplus electricity to pump water back into a high-level reservoir. In normal working conditions the water flows from this reservoir through the ⇨turbines to generate power for feeding into the grid. At times of low power demand, electricity is taken from the grid to turn the turbines into pumps that then pump the water back again. This ensures that there is always a maximum 'head' of water in the reservoir to give the maximum output when required.

An example of a pumped-storage plant is in Dinorwig, North Wales.

pyramid of numbers a diagram that shows how many plants and animals there are at different levels in a ⇨food chain.

There are always far fewer individuals at the top of the chain than at the bottom because only about 10% of the food an animal eats is turned into flesh, so the amount of food flowing through the chain drops at each step. In a pyramid of numbers, the primary producers (usually plants) are represented at the bottom by a broad band, the plant-eaters are shown above by a narrower band, and the animals that prey on them by a narrower band still. At the top of the pyramid are the 'top carnivores' such as lions and sharks, which are present in the smallest number.

pyrolysis decomposition of a substance by heating it to a high temperature in the absence of air. The process is used to burn and dispose of old tyres, for example, without contaminating the atmosphere.

Q

quadrat in environmental studies, a square structure used to study the distribution of plants in a particular place, for instance a field, rocky shore, or mountainside. The size varies, but is usually 0.5 or 1 m square, small enough to be carried easily. The quadrat is placed on the ground and the abundance of species estimated. By making such measurements a reliable understanding of species distribution is obtained.

quetzal long-tailed Central American bird *Pharomachus mocinno* of the trogon family. The male is brightly coloured, with green, red, blue, and white feathers, and is about 1.3 m/4.3 ft long including tail. The female is smaller and lacks the tail and plumage. The quetzal's forest habitat is rapidly being destroyed, and hunting of birds for trophies or souvenirs also threatens its survival.

R

rad unit of absorbed radiation dose, now replaced in the SI system by the ⇨gray (one rad equals 0.01 gray), but still commonly used. It is defined as the dose when one kilogram of matter absorbs 0.01 joule of radiation energy (formerly, as the dose when one gram absorbs 100 ergs).

radiation emission of radiant ⇨energy as particles or waves – for example, heat, light, alpha particles, and beta particles (see ⇨radioactivity). See also ⇨atomic radiation.

radiation heat energy given off by a warm body such as the Sun.

Only a tiny fraction of the Sun's radiation reaches the Earth: as it passes through the ⇨atmosphere much of it is absorbed and scattered by dust, water vapour, and gases such as ozone. The amount that eventually reaches the surface, called insolation, is greatest at the equator, where the Sun is most directly overhead and the radiation most concentrated. At the poles, where the angle of the Sun's rays is much lower, the insolation is less.

Radiation is not only received by the Earth, it is also given off by its surface (***ground radiation***). The climatic difference between continents and oceans is partly due to the fact that land loses and gains heat more rapidly than the sea.

radiation sickness sickness resulting from exposure to radiation, including X-rays, gamma rays, neutrons, and other nuclear radiation, and from weapons and ⇨fallout. Such radiation ionizes atoms in the body and causes nausea, vomiting, diarrhoea, and other symptoms. The body cells themselves may be damaged even by very small doses, causing leukaemia; genetic changes may be induced in the germ plasm, causing infants to be born damaged or mutated.

radiation units units of measurement for radioactivity and radiation doses. Continued use of the units introduced earlier this century (the curie, rad, rem, and roentgen) has been approved while the derived SI units (becquerel, gray, sievert, and coulomb) become familiar. One curie equals 3.7×10^{-10} becquerels (activity); one rad equals 10^{-2} gray (absorbed dose); one rem equals 10^{-2} sievert (dose equivalent); one roentgen equals 2.58×10^{-4} coulomb/kg (exposure to ionizing radiation).

The average radiation exposure per person per year in the USA is one millisievert (0.1 rem), of which 50% is derived from naturally occurring radon.

radioactive decay process of continuous disintegration undergone by the nuclei of radioactive elements, such as radium and various isotopes of uranium and the transuranic elements. This changes the element's atomic number, thus transmuting one element into another, and is accompanied by the emission of radiation. Alpha and beta decay are the most common forms.

In *alpha decay* (the loss of a helium nucleus – two protons and two neutrons) the atomic number decreases by two; in *beta decay* (the loss of an electron) the atomic number increases by one. Certain lighter artificially created isotopes also undergo radioactive decay. The associated radiation consists of alpha rays, beta rays, or gamma rays (or a combination of these), and it takes place at a constant rate expressed as a specific half-life, which is the time taken for half of any mass of that particular isotope to decay completely. Less commonly occurring decay forms include heavy-ion emission, electron capture, and spontaneous fission (in each of these the atomic number decreases).

The original nuclide is known as the parent substance, and the product is a daughter nuclide (which may or may not be radioactive). The final product in all modes of decay is a stable element.

Radioactive Incident Monitoring Network (RIMNET) monitoring network at 46 (to be raised to about 90) Meteorological Office sites throughout the UK. It feeds into a central computer, and was installed 1989 to record contamination levels from nuclear incidents such as the ⇨Chernobyl disaster.

radioactive waste any waste that emits radiation in excess of the background level. See ⇨nuclear waste.

radioactivity spontaneous alteration of the nuclei of radioactive atoms, accompanied by the emission of radiation. It is the property exhibited by the radioactive ⇨isotopes of stable elements and all isotopes of radioactive elements, and can be either natural or induced. See ⇨radioactive decay.

Radioactivity establishes an equilibrium in parts of the nuclei of unstable radioactive substances, ultimately to form a stable arrangement of nucleons (protons and neutrons); that is, a nonradioactive (stable) element. This is most frequently accomplished by the emission of ⇨alpha particles (helium nuclei); ⇨beta particles (electrons and positrons); or ⇨gamma radiation (electromagnetic waves of very high frequency). It takes place either directly, through a one-step decay, or indirectly, through a number of decays that transmute one element into another. This is called a decay series or chain, and sometimes produces an element more radioactive than its predecessor.

radioisotope (contraction of *radioactive* ⇨*isotope*) a naturally occurring or synthetic radioactive form of an element. Most radioisotopes are made by bombarding a stable element with neutrons in the core of a nuclear reactor. The radiations given off by radioisotopes are easy to detect (hence their use as tracers), can in some instances penetrate substantial thicknesses of materials, and have profound effects (such as genetic mutation) on living matter. Although dangerous, radioisotopes are used in the fields of medicine, industry, agriculture, and research.

radium white, radioactive, metallic element, symbol Ra, atomic number 88, relative atomic mass 226.02. It is one of the alkaline-earth metals, found in nature in pitchblende and other uranium ores. Of the 16 isotopes, the commonest, Ra-226, has a half-life of 1.622 years.

Radium decays in successive steps to produce radon (a gas), polonium, and finally a stable isotope of lead. The isotope Ra-223 decays through the uncommon mode of heavy-ion emission, giving off carbon-14 and transmuting directly to lead. Because radium luminesces, it was formerly used in paints that glowed in the dark; when the hazards of radioactivity became known its use was abandoned, but factory and dump sites remain contaminated and many former workers and people living nearby contracted fatal cancers.

radon colourless, odourless, gaseous, radioactive, nonmetallic element, symbol Rn, atomic number 86, relative atomic mass 222. It is grouped with the inert gases and was formerly considered nonreactive, but is now known to form some compounds with fluorine. Of the 20 known isotopes, only three occur in nature; the longest half-life is 3.82 days.

Radon is the densest gas known and occurs in small amounts in spring water, streams, and the air, being formed from the natural radioactive decay of radium. It is responsible for more than half of the annual average dose of radiation. The danger of radon stems from the way its atoms break down into radioactive particles which can be inhaled and deposited in the lungs.

The average radon radiation level found in a study of 40 British limestone caves was 2,900 becquerels per cubic metre. This compares with the National Radiological Protection Board's set level of 200 becquerels per cubic metre, at which removal of radon from homes is recommended. The highest levels were found in the Giant's Hole in Derbyshire, with values of around 155,000 becquerels per cubic metre during the summer, which is the highest level ever recorded from a natural limestone cave. This compares with a maximum level of 54,000 becquerels per cubic metre found in a limestone cave in the USA.

railway method of transport in which trains convey passengers and goods along a twin rail track. Following the work of British steam pioneers such as Scottish engineer James Watt, English engineer George Stephenson built the first public steam railway, from Stockton to Darlington, England, in 1825. This heralded extensive railway building in Britain, continental Europe, and North America, providing a fast and economical means of transport and communication. After World War II, steam engines were replaced by electric and diesel engines. At the same time, the growth of road building, air services, and car ownership destroyed the supremacy of the railways.

There are many environmental advantages to be gained by maximum rail use. Railways are relatively safe and nonpolluting, have low space requirements, are energy-efficient, fast and could be accessible to everyone, unlike cars which can only be used by those with sufficient skills and financial resources.

rainforest dense forest usually found on or near the equator where the climate is hot and wet. Heavy rainfall results as the moist air brought by the converging tradewinds rises because of the heat. Over half the tropical rainforests are in Central and South America, the rest in SE Asia and Africa. They provide the bulk of the oxygen needed for plant and animal respiration. Tropical rainforest once covered 14% of the Earth's land surface, but are now being destroyed at an increasing rate as their valuable timber is harvested and the land cleared for agriculture, causing problems of ⇨deforestation. In the last 30 years, Central America has lost almost two-thirds of its rainforests to cattle ranching. Although by 1991 over 50% of the world's rainforest had been removed, they still comprise about 50% of all growing wood on the planet, and harbour at least 40% of the Earth's species (plants and animals).

Tropical rainforests are characterized by a great diversity of species, usually of tall broad-leaved evergreen trees, with many climbing vines and ferns, some of which are a main source of raw materials for medicines. A tropical forest, if properly preserved, can yield medicinal plants, oils (from cedar, juniper, cinnamon, sandalwood), spices, gums, resins (used in inks, lacquers, linoleum), tanning and dyeing materials, forage for animals, beverages, poisons, green manure, rubber, and animal products (feathers, hides, honey). Paradoxically, although tropical rainforests are biologically rich, their soils are extremely poor as any nutrients in the soil are immediately absorbed by the efficient rooting system of the tropical trees. Other rainforests include montane, upper montane or cloud, mangrove, and subtropical.

Rainforests comprise some of the most complex and diverse ecosystems on the planet and help to regulate global weather patterns. When deforestation occurs, the microclimate of the mature forest disappears; soil erosion and flooding become major problems since rainforests protect the shallow tropical soils. Once an area is cleared it is very difficult for shrubs and bushes to re-establish because nutrients are minimal. This has caused problems for plans to convert rainforests into agricultural land – after two or three years the crops fail and the land is left bare. Clearing of the rainforests may lead to a global warming of the atmosphere, and contribute to the ⇨greenhouse effect.

recycling processing of industrial and household waste (such as paper, glass, and some metals and plastics) so that it can be reused. This saves expenditure on scarce raw materials, slows down the depletion of ⇨nonrenewable resources, and helps to reduce pollution.

The USA recycles only around 13% of its waste, compared to around 33% in Japan. However, all US states encourage or require local recycling programmes to be set up. It was estimated 1992 that 4,000 cities collected waste from 71 million people for recycling. Most of these programmes were set up in 1989–92. Around 33% of newspapers, 22% of office paper, 64% of aluminium cans, 3% of plastic containers, and 20% of all glass bottles and jars were recycled. Aluminium is frequently recycled because of its value and special properties which allow it to be melted down and re-pressed without loss of quality, unlike paper and glass which deteriorate when recycled.

Most British recycling schemes are voluntary, and rely on people taking waste items to a central collection point. However, some local authorities, such as Leeds, now ask householders to separate waste before collection, making recycling possible on a much larger scale.

refining any process that purifies or converts something into a more useful form. Metals usually need refining after they have been extracted from their ores by such processes as smelting. Petroleum, or crude oil, needs refining before it can be used; the process involves fractional distillation, the separation of the substance into separate components or 'fractions'.

Electrolytic metal-refining methods use the principle of electrolysis to obtain pure metals. When refining petroleum, or crude oil, further refinery processes after fractionation convert the heavier fractions into more useful lighter products. The most important of these processes is cracking; others include polymerization, hydrogenation, and reforming.

relative biological effectiveness (RBE) the relative damage caused to living tissue by different types of radiation. Some radiations do much more damage than others; alpha particles, for example, cause

20 times as much destruction as electrons (beta particles).

The RBE is defined as the ratio of the absorbed dose of a standard amount of radiation to the absorbed dose of 200 kV X-rays that produces the same amount of biological damage.

rem acronym of *roentgen equivalent man* SI unit of radiation dose equivalent. The rem has now been replaced in the SI system by the ⇨sievert (one rem equals 0.01 sievert), but remains in common use.

renewable energy power from any source that replenishes itself. Most renewable systems rely on ⇨solar energy directly or through the weather cycle as ⇨wave power, ⇨hydroelectric power, or wind power via ⇨wind turbines, or solar energy collected by plants (alcohol fuels, for example). In addition, the gravitational force of the Moon can be harnessed through ⇨tidal power stations, and the heat trapped in the centre of the Earth is used via ⇨geothermal energy systems.

renewable resource natural resource that is replaced by natural processes in a reasonable amount of time. Soil, water, forests, plants, and animals are all renewable resources as long as they are properly conserved. Fossil fuels are not considered renewable because they take millions of years to form. Solar, wind, wave, and geothermal energies are based on renewable resources.

respiration biochemical process whereby food molecules are progressively broken down (oxidized) to release energy in the form of ATP (adenosine triphosphate). In most organisms this requires oxygen, but in some bacteria the oxidant is the nitrate or sulphate ion instead. In all higher organisms, respiration occurs in the mitochondria. Respiration is also used to mean breathing, although this is more accurately described as a form of gas exchange.

reuse multiple use of a product (often a form of packaging), by returning it to the manufacturer or processor each time. Many such returnable items are sold with a deposit which is reimbursed if the item is returned. Reuse is usually more energy- and resource-efficient than ⇨recycling unless there are large transport or cleaning costs. One of the most notable examples is the milk bottle in the UK,

rhinoceros odd-toed hoofed mammal of the family Rhinocerotidae. The one-horned Indian rhinoceros *Rhinoceros unicornis* is up to 2 m/

6 ft high at the shoulder, with a tubercled skin, folded into shieldlike pieces; the African rhinoceroses are smooth-skinned and two-horned.

Needless slaughter has led to the near extinction of all species of rhinoceros, particularly the Sumatran rhinoceros and ⇨Javan rhinoceros. In Zimbabwe, the government introduced a two-month programme of dehorning rhinos to make them worthless to poachers. A 4 kg/9 lb rhino horn may fetch up to $58,000. In 1991, poachers killed more than 1,000 rhinos and there are now only about 3,000 black rhino left in Africa, about 1,000 of them in Zimbabwe. In 1991, the total world population was 11,000, only about half the number considered safe for the survival of the species.

roentgen or *röntgen* unit (symbol R) of radiation exposure, used for X-rays and gamma rays. It is defined in terms of the number of ions produced in one cubic centimetre of air by the radiation. Exposure to 1,000 roentgens gives rise to an absorbed dose of about 870 rads (8.7 grays), which is a dose equivalent of 870 rems (8.7 sieverts).

The annual dose equivalent from natural sources in the UK is 1,100 microsieverts.

Romania *although sulphur-dioxide levels are low, the country faces serious environmental problems from industrial pollution. Only 20% of the country's rivers can provide drinkable water as a result of pollution and the haphazard dumping of both industrial and domestic waste has led to land degradation and health problems in much of the country.*

---

röntgen alternative spelling for ⇨roentgen, unit of X- and gamma-ray exposure.

rural depopulation loss of people from remote country areas to cities; it is an effect of migration. In poor countries, large-scale migration to urban core regions may deplete the countryside of resources and workers. The population left behind will be increasingly aged, and agriculture declines.

In parts of the UK (for example, East Anglia), mechanization of agriculture in the 1950s and 1960s has led to the push factor of declining job prospects, and especially the young have moved away. Rural depopulation may result in reduced village services.

RUSSIAN NUCLEAR POWER STATIONS CAUSE CONCERN

Russian officials played down the seriousness of a leak of radioactive iodine from a Chernobyl-type nuclear reactor near St Petersburg in 1992. Minatom, Russia's atomic energy ministry, said 'Releases of inert gases and iodine to the environment do not exceed the requirements of sanitary rules and regulation.' Nevertheless, the incident sparked fears of another Chernobyl-type disaster. There is a growing consensus that Russia's nuclear technology is out of date and dangerous.

The incident took place at Sosnovy Bor, 100 km/60 mi west of St Petersburg. It is the home of four, 16-year-old, RMBK reactors. In these reactors, the fuel channels form a honeycomb structure within the graphite core through which cooling water flows; damage to the core in one of the reactors caused the accident. After Chernobyl, all RMBK reactors were fitted with extra safety systems. At Sosnovy Bor, the emergency control systems appeared to work correctly, shutting down the reactor immediately. Nevertheless, a week later a government ecology adviser claimed the accident had released 109 times more radiation than had been officially acknowledged. A group of Western nuclear experts visited the Sosnovy Bor site and were disturbed by the high levels of radiation in the plant and the old-fashioned control technology. This prompted the authorities in Ukraine to shut down the remaining reactors at Chernobyl for maintenance and safety checks.

Health risks

Two reactors were recently shut down in the central Siberian city of Krasnoyarsk. The reactors were said to be a danger to civilian health and the environment. The old-fashioned reactors were cooled by water taken from and pumped back into the Yenisey River. Pollution has already made water from 75% of Russia's rivers, lakes and reservoirs unfit to drink. The local population feared that plutonium, a dangerous poison from the reactors, would find its way to the river.

The emerging pattern is one of old-fashioned and unsafe technology. Unfortunately, the solution to the problem is far from easy. The country relies on nuclear power. Its 45 nuclear plants produce well over 200 billion units of electricity each year.

Russian Federation *the country's enormous array of natural resources is matched by the legacy of environmental problems left by the Soviet era of heavy industrialization with little regard for environmental impact. Much of the country suffers from serious pollution of the air, land, and waterways from both industrial and agricultural sources. The area surrounding Chelyabinsk, in the Urals, where a large plutonium plant is stationed is reckoned to be one of the most radioactive in the world.*

S

sable marten *Martes zibellina*, about 50 cm/20 in long and usually brown. It is native to N Eurasian forests, but now found mainly in E Siberia. The sable has diminished in numbers because of its valuable fur, which has long attracted hunters. Conservation measures and sable farming have been introduced to save it from extinction.

salinization the accumulation of salt in water or soil; it is a factor in ⇨desertification.

Sandoz pharmaceutical company whose plant in Basel, Switzerland, suffered an environmentally disastrous chemical fire in Nov 1986. Hundreds of tonnes of pesticides, including mercury-based fungicides, spilled into the river Rhine, rendering it lifeless for 100–200 km/60–120 mi and killing about half a million fish.

Saudi Arabia oil pollution caused by the Gulf War 1990–91 has affected 460 km/285 mi of the Saudi coastline, threatening desalination plants and damaging the wildlife of salt marshes, mangrove forest, and mudflats. As in many countries in the region, water supply is a serious problem – irrigation has seriously depleted the ⇨aquifers as water is pumped out faster than it can be replenished.

Sellafield site of a nuclear power station on the coast of Cumbria, NW England. It was known as **Windscale** until 1971, when the management of the site was transferred from the UK Atomic Energy Authority to British Nuclear Fuels Ltd. The plant is the world's greatest discharger of radioactive waste: between 1968 and 1979 180 kg/400 lb of plutonium was discharged into the Irish Sea. In 1990 a study revealed an increased risk of leukaemia in children whose fathers worked at Sellafield 1950–85. For accidents, see ⇨nuclear safety.

sere plant ⇨succession developing in a particular habitat. A *lithosere* is a succession starting on the surface of bare rock. A *hydrosere* is a succession in shallow freshwater, beginning with planktonic vegetation and the growth of pondweeds and other aquatic plants, and ending with the development of swamp. A *plagiosere* is the sequence of communities that follows the clearing of the existing vegetation.

sewage disposal the disposal of human excreta and other water-borne waste products from houses, streets, and factories. Conveyed through sewers to sewage works, sewage has to undergo a series of treatments to be acceptable for discharge into rivers or the sea, according to various local laws and ordinances. Raw sewage, or sewage that has not been treated adequately, is a serious source of water pollution and a cause of ⇨eutrophication.

In the industrialized countries of the West, most industries are responsible for disposing of their own wastes. Government agencies establish industrial waste-disposal standards. In most countries, sewage works for residential areas are the responsibility of local authorities. The solid waste (sludge) may be spread over fields as a fertilizer or, in a few countries, dumped at sea. A significant proportion of bathing beaches in densely populated regions have an unacceptably high bacterial content, largely as a result of untreated sewage being discharged into rivers and the sea. This can, for example, cause stomach upsets in swimmers.

The use of raw sewage as a fertilizer (long practised in China) has the drawback that disease-causing microorganisms can survive in the soil and be transferred to people or animals by consumption of subsequent crops. Sewage sludge is safer, but may contain dangerous levels of heavy metals and other industrial contaminants.

In 1987, Britain dumped more than 4,700 tonnes of sewage sludge into the North Sea, and 4,200 tonnes into the Irish Sea and other coastal areas. Also dumped in British coastal waters, other than the Irish Sea, were 6,462 tonnes of zinc, 2,887 tonnes of lead, 1,306 tonnes of chromium, and 8 tonnes of arsenic. Some 916 tonnes of zinc, 297 tonnes of lead, 200 tonnes of chromium, and 1 tonne of arsenic were dumped into the Irish Sea. Strict European rules will phase out sea dumping by 1998.

shark any member of various orders of cartilaginous fishes (class Chondrichthyes), found throughout the oceans of the world. There are about 400 known species of shark. They have tough, usually grey, skin covered in denticles (small toothlike scales). A shark's streamlined body has side pectoral fins, a high dorsal fin, and a forked tail with a large upper lobe. Five open gill slits are visible on each side of the generally pointed head.

Game fishing for 'sport', the eradication of sharks in swimming and recreation areas, and their industrial exploitation as a source of leather, oil, and protein have reduced their numbers. Some species, such as the great white shark, the tiger shark, and the hammerhead, are now endangered and their killing has been banned in US waters since July 1991. Other species will be protected by catch quotas.

shifting cultivation farming system where farmers move on from one place to another. The most common form is slash-and-burn agriculture: land is cleared by burning, so that crops can be grown. After a few years, soil fertility is reduced and the land is abandoned. A new area is cleared while the old land recovers its fertility.

Slash-and-burn is practised in many tropical forest areas, such as the Amazon region, where yams, manioc, and sweet potatoes can be grown. This system works well while population levels are low, but where there is ⇨overpopulation, the old land will be reused before soil fertility has been restored. A variation of this system, found in parts of Africa, is rotational bush fallowing.

sick building syndrome malaise diagnosed in the early 1980s among office workers and thought to be caused by such pollutants as formaldehyde (from furniture and insulating materials), benzene (from paint), and the solvent trichloroethene, concentrated in air-conditioned buildings. Symptoms include headache, sore throat, tiredness, colds, and flu. Studies have found that it can cause a 40% drop in productivity and a 30% rise in absenteeism.

Work on improving living conditions for astronauts showed that the causes were easily removed by potplants in which interaction is thought to take place between the plant and microorganisms in its roots. Among the most useful are chrysanthemums (counteracting benzene), English ivy and the peace lily (trichloroethene), and the spider plant (formaldehyde).

sievert SI unit (symbol Sv) of radiation dose equivalent. It replaces the rem (1 Sv equals 100 rem). Some types of radiation do more damage than others for the same absorbed dose – for example, the same absorbed dose of alpha radiation causes 20 times as much biological damage as the same dose of beta radiation. The equivalent dose in sieverts is equal to the absorbed dose of radiation in rays multiplied by the relative biological effectiveness. Humans can absorb up to 0.25 Sv without immediate ill effects; 1 Sv may produce radiation sickness; and more than 8 Sv causes death.

Single European Act 1986 update of the Treaty of Rome (signed 1957) that provides a legal basis for action by the European Community in matters relating to the environment. The act requires that environmental protection shall be a part of all other Community policies. Also, it allows for agreement by a qualified majority on some legislation, whereas before such decisions had to be unanimous.

site of special scientific interest (SSSI) in the UK, land that has been identified as having animals, plants, or geological features that need to be protected and conserved. From 1991 these sites were designated and administered by English Nature, Scottish Natural Heritage, and the Countryside Council for Wales.

Numbers fluctuate, but there were over 5,000 SSSIs 1991, covering about 6% of Britain. Although SSSIs enjoy some legal protection, this does not in practice always prevent damage or destruction; during 1992, for example, 40% of SSSIs were damaged by development, farming, public access and neglect. A report by English Nature estimated a quarter of the total area of SSSI's, over 1 million acres, had been damaged by acid rain. Around 1% of SSSIs are irreparably damaged each year.

slag the molten mass of impurities that is produced in the smelting or refining of metals. The slag produced in the manufacture of iron in a blast furnace contains mostly silicates, phosphates, and sulphates of calcium. When cooled, the solid is broken up and used as a core material in the foundations of roads and buildings.

slash and burn simple agricultural method whereby the natural vegetation is cut and burned, and the clearing then farmed for a few years until the soil loses its fertility, whereupon farmers move on and leave the area to regrow.

Although this is possible with a small, widely dispersed population, it becomes unsustainable with more people and is now a form of ⇨deforestation. Slash and burn is particularly inappropriate in tropical rainforest areas because the soil is already poor and once the trees are removed it becomes still further impoverished.

slurry form of manure composed mainly of liquids. Slurry is collected and stored on many farms, especially when large numbers of animals are kept in factory units (see ⇨factory farming). When slurry tanks are accidentally or deliberately breached, large amounts can spill into rivers, killing fish and causing ⇨eutrophication. Some slurry is spread on fields as a fertilizer.

Slurry spills in the UK increased enormously in the 1980s and were by 1991 running at rates of over 4,000 a year. Tighter regulations were being introduced to curb this pollution.

Small is Beautiful book by German economist E F Schumacher (1911–1973), published 1973, which argues that the increasing scale of corporations and institutions, concentration of power in fewer hands, and the overwhelming priority being given to economic growth are both unsustainable and disastrous to environment and society.

smog natural fog containing impurities (unburned carbon and sulphur dioxide) from domestic fires, industrial furnaces, certain power stations, and ⇨internal-combustion engines (petrol or diesel). It can cause substantial illness and loss of life, particularly among chronic bronchitics, and damage to wildlife.

The London smog of 1952 killed 4,000 people from heart and lung diseases. The use of smokeless fuels, the treatment of effluent, and penalties for excessive smoke from poorly maintained and operated vehicles can be extremely effective in cutting down smog, as in London, but it still occurs in many cities throughout the world.

smokeless fuel fuel that does not give off any smoke when burned, because all the carbon is fully oxidized to carbon dioxide (CO_2). Natural gas, oil, and coke are smokeless fuels.

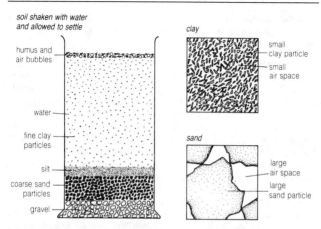

soil shaken with water
and allowed to settle

humus and
air bubbles

water

fine clay
particles

silt

coarse sand
particles

gravel

clay

small
clay particle

small
air space

sand

large
air space

large
sand particle

soil *The constituents of a soil sample are revealed if the soil is shaken with water and allowed to settle. The mineral nature of soil is determined by the nature of the weathered rock or sediment. Climate often determines the texture and size of the particles and the degree of consolidation.*

soil loose covering of broken rocky material and decaying organic matter overlying the bedrock of the Earth's surface. Various types of soil develop under different conditions: deep soils form in warm wet climates and in valleys; shallow soils form in cool dry areas and on slopes. *Pedology*, the study of soil, is significant because of the relative importance of different soil types to agriculture.

The organic content of soil is widely variable, ranging from zero in some desert soils to almost 100% in peats. Soils influence the type of agriculture employed in a particular region – light well-drained soils favour arable farming, whereas heavy clay soils give rise to lush pasture land.

Soil Association pioneer British ecological organization founded 1946, which campaigns against pesticides and promotes organic farming.

soil depletion decrease in soil quality over time. Causes include loss of nutrients caused by overfarming, erosion by wind, and chemical imbalances caused by acid rain.

soil erosion the wearing away and redistribution of the Earth's soil layer. It is caused by the action of water, wind, and ice, and also by improper methods of ⇨agriculture. If unchecked, soil erosion results in the formation of deserts (⇨desertification). It has been estimated that 20% of the world's cultivated topsoil was lost between 1950 and 1990.

If the rate of erosion exceeds the rate of soil formation (from rock and decomposing organic matter), then the land will decline and eventually become infertile. The removal of forests (⇨deforestation) or other vegetation often leads to serious soil erosion, because plant roots bind soil, and without them the soil is free to wash or blow away, as in the American ⇨dust bowl. The effect is worse on hillsides, and there has been devastating loss of soil where forests have been cleared from mountainsides, as in Madagascar.

Improved agricultural practices such as contour ploughing are needed to combat soil erosion. Windbreaks, such as hedges or strips planted with coarse grass, are valuable, and organic farming can reduce soil erosion by as much as 75%.

Soil degradation and erosion are becoming as serious as the loss of the rainforest. It is estimated that more than 10% of the world's soil lost a large amount of its natural fertility during the latter half of the 20th century. Some of the worst losses are in Europe, where 17% of the soil is damaged by human activity such as mechanized farming and fallout from acid rain. Mexico and Central America have 24% of soil highly degraded, mostly as a result of deforestation.

solar energy energy derived from the Sun's radiation. The amount of energy falling on just 1 sq km/0.4 sq mi is about 4,000 megawatts, enough to heat and light a small town. In one second the Sun gives off 13 million times more energy than all the electricity used in the USA in one year. *Solar heaters* have industrial or domestic uses. They usually consist of a black (heat-absorbing) panel containing pipes through which air or water, heated by the Sun, is circulated, either by thermal convection or by a pump.

Solar energy may also be harnessed indirectly using *solar cells* (photovoltaic cells) made of panels of semiconductor material (usually silicon), which generate electricity when illuminated by sunlight. Although it is difficult to generate a high output from solar energy compared to sources such as nuclear or fossil fuels, it is a major nonpolluting and renewable energy source used as far north as Scandinavia as well as in the SW USA and in Mediterranean countries.

A solar furnace, such as that built in 1970 at Odeillo in the French Pyrénées, has thousands of mirrors to focus the Sun's rays; it produces uncontaminated intensive heat (up to 3,000°C/5,4000°F) for industrial and scientific or experimental purposes. The world's first solar power station connected to a national grid opened 1991 at Adrano in Sicily. Scores of giant mirrors move to follow the Sun throughout the day, focusing the rays into a boiler. Steam from the boiler drives a conventional turbine. The plant generates up to 1 megawatt. A similar system, called Solar 1, has been built in the Mojave desert near Daggett, California, USA. It consists of 1,818 computer-controlled mirrors arranged in circles around a central boiler tower 91 m/300 ft high. Advanced schemes have been proposed that would use giant solar reflectors in space to harness solar energy and beam it down to Earth in the form of microwaves. Despite their low running costs, their high installation cost and low power output have meant that solar cells have found few applications outside space probes and artificial satellites. Solar heating is, however, widely used for domestic purposes in many parts of the world, and is an important renewable source of energy.

solar pond natural or artificial 'pond', such as the Dead Sea, in which salt becomes more soluble in the Sun's heat. Water at the bottom becomes saltier and hotter, and is insulated by the less salty water layer at the top. Temperatures at the bottom reach about 100°C/212°F and can be used to generate electricity.

solenodon rare insectivorous shrewlike mammal, genus *Solenodon*. There are two species, one each on Cuba and Hispaniola, and they are threatened with extinction owing to introduced predators. They are about 30 cm/1 ft long with a 25 cm/10 in tail, slow-moving, and they produce venomous saliva.

solvent substance, usually a liquid, that will dissolve another substance. Although the commonest solvent is water, in popular use the term refers to low-boiling-point organic liquids, which are harmful if used in a confined space. They can give rise to respiratory problems, liver damage, and neurological complaints.

Typical organic solvents are petroleum distillates (in glues), xylol (in paints), alcohols (for synthetic and natural resins such as shellac), esters (in lacquers, including nail varnish), ketones (in cellulose lacquers and resins), and chlorinated hydrocarbons (as paint stripper and dry-cleaning fluids). The fumes of some solvents, when inhaled (glue-sniffing), affect mood and perception. In addition to damaging the brain and lungs, repeated inhalation of solvent from a plastic bag can cause death by asphyxia.

Somalia *destruction of trees for fuel and by grazing livestock has led to an increase in desertification. Civil war has exacerbated Somalia's perennial affliction with drought – two out of every five years are expected to be drought years. Recently the problem has been worsened still further by excessive numbers of livestock consuming valuable water resources from wells to provide year-round grazing.*

sonic boom noise like a thunderclap that occurs when an aircraft passes through the sound barrier, or begins to travel faster than the speed of sound. It happens when the cone-shaped shock wave caused by the plane touches the ground.

SSSI abbreviation for ⇨*Site of Special Scientific Interest*.

standing crop the total number of individuals of a given species alive in a particular area at any moment. It is sometimes measured as the weight (or ⇨biomass) of a given species in a sample section.

sterilization any surgical operation to terminate the possibility of reproduction. In women, this is normally achieved by sealing or tying off the Fallopian tubes (tubal ligation) so that fertilization can no longer take place. In men, the transmission of sperm is blocked by vasectomy.

Sterilization is a safe alternative to ⇨contraception and may be encouraged by governments to limit population growth or as part of a selective-breeding policy (see ⇨eugenics).

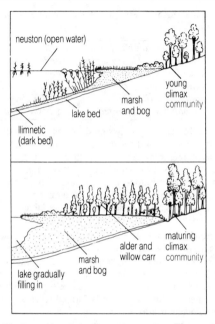

succession *The succession of plant types along a lake. As the lake gradually fills in, a mature climax community of trees forms inland from the shore. Extending out from the shore, a series of plant communities can be discerned with small, rapidly growing species closest to the shore.*

succession a series of changes that occur in the structure and composition of the vegetation in a given area from the time it is first colonized by plants (***primary succession***), or after it has been disturbed by fire, flood, or clearing (***secondary succession***).

If allowed to proceed undisturbed, succession leads naturally to a stable ⇨climax community (for example, oak and hickory forest or savanna grassland) that is determined by the climate and soil characteristics of the area.

Sudan rapid desertification has led to substantial losses of agricultural land and some regions being entirely abandoned as they will no longer sustain the local population. Contaminated water and inadequate public health resources have led to widespread epidemics and outbreaks of malaria are rife. The building of the Jonglei canal to supply water to N Sudan and Egypt threatens the grasslands of S Sudan.

sulphur dioxide SO_2 pungent gas produced by burning sulphur in air or oxygen. It is widely used for disinfecting food vessels and equipment, and as a preservative in some food products. It occurs in industrial flue gases and is a major cause of ⇨acid rain.

Although UK SO_2 emissions have decreased from 7 million tonnes 1970 to 4 million tonnes 1990, SO_2 remains an important contributory factor in acid rain. It is also a serious health hazard to children, the elderly, and those with respiratory problems.

sulphuric acid or *oil of vitriol* H_2SO_4 a dense, viscous, colourless liquid that is extremely corrosive. It gives out heat when added to water and can cause severe burns. Sulphuric acid is used extensively in the chemical industry, in the refining of petrol, and in the manufacture of fertilizers, detergents, explosives, and dyes. It forms the acid component of car batteries. Sulphuric acid is formed when sulphur dioxide dissolves in rainwater to form ⇨acid rain.

sustainable capable of being continued indefinitely. For example, the sustainable yield of a forest is equivalent to the amount that grows back. Environmentalists made the term a catchword, in advocating the sustainable use of resources.

sustained-yield cropping the removal of surplus individuals from a ⇨population of organisms so that the population maintains a constant size. This usually requires selective removal of animals of all ages and both sexes to ensure a balanced population structure. Taking too many individuals can result in a population decline, as in ⇨overfishing. Excessive cropping of young females may lead to fewer births in following years, and a fall in population size. Appropriate cropping frequencies can be determined from an analysis of a life table.

Sweden *like Norway, Sweden suffers badly from acid rain of foreign origin. Despite having made great efforts to reduce its own emissions of sulphur dioxide, this accounts for only 12% of the country's annual acid rain, the rest coming from neighbouring countries. Of the country's 90,000 lakes, 20,000 are affected by acid rain; 4,000 are so severely acidified that no fish are thought to survive in them.*

---

sweetener any chemical that gives sweetness to food. Caloric sweeteners are various forms of sugar; noncaloric, or artificial, sweeteners are used by dieters and diabetics and provide neither energy nor bulk. Questions have been raised about the long-term health effects from several artificial sweeteners.

Sweeteners are used to make highly processed foods attractive, whether sweet or savoury. Most of the noncaloric sweeteners do not have ⇨E numbers. Some are banned from baby foods and for young children: thaumatin, aspartame, acesulfame-K, sorbitol, and mannitol. Cyclamate is banned in the UK and the USA; acesulfame-K is banned in the USA.

Switzerland *an estimated 43% of coniferous trees, particularly in the central Alpine region, have been killed by acid rain, 90% of which comes from other countries. Over 50% of bird species are classified as threatened due to habitat destruction and pollution. Switzerland has made great efforts to tackle its environmental problems, with vehicle emission limits set well below those of the rest of Europe and the import or production of leaded petrol banned. About 90% of the municipal population is served by treatment plants for waste water.*

T

taiga or *boreal forest* Russian name for the forest zone south of the ⇨tundra, found across the northern hemisphere. Here, dense forests of conifers (spruces and hemlocks), birches, and poplars occupy glaciated regions punctuated with cold lakes, streams, bogs, and marshes. Winters are prolonged and very cold, but the summer is warm enough to promote dense growth.

The varied fauna and flora are in delicate balance because the conditions of life are so precarious. This ecology is threatened by mining, forestry, and pipeline construction.

Taiwan industrialization has taken its toll: an estimated 30% of the annual rice crop is dangerously contaminated with mercury, cadmium, and other heavy metals. Taiwan has been condemned by environmentalists for its continued involvement in the illegal trade in rhino horn, ivory, tiger products, and other endangered species.

Tanzania deforestation and land degradation due to unsustainable agricultural methods are major concerns; if forest continues to be lost at the present rate, the country will have no forest left within a century. The black rhino faces extinction as a result of poaching; the use of dynamite by fishermen is damaging the country's coral reef which is an important factor for tourism as well as the survival of the fishery.

tartrazine (E102) yellow food colouring produced synthetically from petroleum. Many people are allergic to foods containing it. Typical effects are skin disorders and respiratory problems. It has been shown to have an adverse effect on hyperactive children.

TBT (abbreviation for ⇨*tributyl tin*) chemical used in antifouling paints that has become an environmental pollutant.

terrapin member of some species of the order Chelonia (⇨turtles and ⇨tortoises). Terrapins are small to medium-sized, aquatic or semi-aquatic, and are found widely in temperate zones. They are omnivorous, but generally eat aquatic animals. Some species are in danger of extinction owing to collection for the pet trade; most of the animals collected die in transit.

tetrachloromethane CCl_4 or *carbon tetrachloride* chlorinated organic compound that is a very efficient solvent for fats and greases, and was at one time the main constituent of household dry-cleaning fluids and of fire extinguishers used with electrical and petrol fires. Its use became restricted after it was discovered to be carcinogenic and it has now been largely removed from educational and industrial laboratories.

tetraethyl lead $Pb(C_2H_5)_4$ compound added to leaded petrol as a component of antiknock to increase the efficiency of combustion in car engines. It is a colourless liquid that is insoluble in water but soluble in organic solvents such as benzene, ethanol, and petrol.

textured vegetable protein manufactured meat substitute; see ⇨TVP.

Thermal Oxide Reprocessing Plant (THORP) nuclear plant built at Sellafield for reprocessing spent fuel from countries around the world, with plutonium as the end product. Court action by Greenpeace attempted to stop THORP from starting up, but failed early 1994.

Third World term originally applied collectively to those countries of Africa, Asia, and Latin America that were not aligned with either the Western bloc (First World) or Communist bloc (Second World). The term later took on economic connotations and was applied to the countries of the ⇨developing world.

THORP acronym for ⇨Thermal Oxide Reprocessing Plant.

Three Mile Island island in the Shenandoah River near Harrisburg, Pennsylvania, USA, site of a nuclear power station which was put out of action following a major accident March 1979. Opposition to nuclear power in the USA was reinforced after this accident and safety standards reassessed. See ⇨nuclear safety.

tidal energy energy derived from the tides. The tides mainly gain their potential energy from the gravitational forces acting between the Earth and the Moon. If water is trapped at a high level during high tide, perhaps by means of a barrage across an estuary, it may then be gradually released and its associated gravitational potential energy exploited to drive turbines and generate electricity. Several schemes have been proposed for the Bristol Channel, in SW England, but environmental concerns as well as construction costs have so far prevented any decision from being taken.

tidal power station ⇨hydroelectric power plant that uses the 'head' of water created by the rise and fall of the ocean tides to spin the water turbines. The world's only large tidal power station is located on the estuary of the river Rance in the Gulf of St Malo, Brittany, France, which has been in use since 1966. It produces 240 megawatts and can generate electricity on both the ebb and flow of the tide.

tiger largest of the great cats *Panthera tigris*, formerly found in much of central and S Asia but nearing extinction because of hunting and the destruction of its natural habitat. The tiger can grow to 3.6 m/12 ft long and weigh 300 kg/660 lbs; it has a yellow-orange coat with black stripes. It is solitary, and feeds on large ruminants. It is a good swimmer.

   The largest of the species is the Siberian tiger; only about 250 remained in the wild 1993. In Sumatra there are only about 800 tigers left, and 200 are killed each year by poachers. They are continually losing their jungle habitat: companies plunder the forest for timber and minerals, and then farmers, often transplanted from other parts of Indonesia, move in and take over the land, often ruining it after a few years owing to poor farming practices. In 1992, the world's tiger population was estimated at 6,000–9,000, of which half were Bengal tigers.

Times Beach town in Missouri, USA, that accidentally became contaminated with ⇨dioxin, and was bought by the Environmental Protection Agency 1983 for cleansing.

tokamak experimental machine designed by Soviet scientists to investigate controlled nuclear fusion. It consists of a doughnut-shaped chamber surrounded by electromagnets capable of exerting very powerful magnetic fields. The fields are generated to confine a very hot (millions of degrees) plasma of ions and electrons, keeping it away from the chamber walls. See also ⇨JET.

topsoil the upper, cultivated layer of soil, which may vary in depth from 8–45 cm(3–18 in). It contains organic matter – the decayed remains of vegetation, which plants need for active growth – along with a variety of soil organisms, including earthworms.

tortoise reptile of the order Chelonia, family Testudinidae, with the body enclosed in a hard shell. Tortoises are related to the ⇨terrapins and ⇨turtles, and range in length from 10 cm/4 in to 150 cm/5 ft.

Many species are under threat due to hunting for sale as pets in the West or for food or other products. Best known in the pet trade is the small *spur-thighed tortoise Testudo graeca*, found in Asia Minor, the Balkans, and N Africa. It was extensively exported, often in appalling conditions, until the 1980s, when strict regulations were introduced to prevent its probable extinction. The *giant tortoises* of the Galápagos and Seychelles may reach a length of 150 cm/5 ft and weigh over 225 kg/500 lbs, and can yield about 90 kg/200 lbs of meat; hence its almost complete extermination by sailors in passing ships. *Tortoiseshell* is in fact the semitransparent shell of the hawksbill ⇨turtle.

tourism visit to a place away from home that (unlike recreation) involves at least an overnight stay. An area that attracts tourism can increase its wealth and job opportunities, although the jobs may be low paid and seasonal. Among the negative effects of tourism are traffic congestion and damage to the environment.

In the UK, tourism generates £23 billion a year and accounts for 4% of gross domestic product (GDP). It provides jobs for about 1.4 million people.

toxicity tests tests carried out on new drugs, cosmetics, food additives, pesticides, and other synthetic chemicals to see whether they are safe for humans to use. They aim to identify potential toxins, carcinogens, teratogens, and mutagens.

Traditionally such tests use live animals such as rats, rabbits, and mice. Examples include the LD50 test, in which increasingly higher dosages of a substance are force fed to a sample group of animals until

1993: YEAR OF THE ROAD PROTESTS

'1993 has been the year of the roads protester' declared the *Independent* newspaper in round-up of events at the end of a year that had seen the anti-roads movement in the UK rise from the ashes of apparent defeat to become a standard bearer of environmental change.

Twyford Down

In December 1992, the Department of Transport seemed finally to have won its twenty year battle to extend the M3 motorway through Twyford Down, near Winchester, Hampshire. Private security guards had evicted the self-styled 'Dongas tribe' – a collective of New Age travellers and radical environmentalists who formed the last rump of protesters – from the site and brought in the bulldozers. However, the DOT's heavy-handed tactics coupled with the Dongas' startling idealism served mainly to galvanize opinion and another camp was quickly started near the site by the Dongas, Earth First!, and other radical protesters. On 6 March, about 400 protesters invaded the site and halted all work for the day – the first of many demonstrations which were to take place at the site almost daily for the next three months. Typically, the protesters would leap onto the working earthmovers and chain themselves by their necks to the hydraulics – a tactic that was not only highly effective but had strong visual impact for the media. Protesters earned considerable respect for their scrupulous nonviolence, in stark contrast with the often brutal heavy-handedness of private security guards employed by the DOT to remove them.

The Twyford Down demonstrations only ended when the DOT gained an injunction on the site and seven protesters were jailed in July for breaching it. Even now, the tactic backfired on the DOT, as the protesters were viewed as martyrs by environmentalists and much of the media. Even the judge who imprisoned them noted that 'civil disobedience is an honourable tradition' and predicted that they would be 'vindicated by history'.

Success in Oxleas

By now it was clear that no amount of protest could save Twyford Down but an example had been given and the momentum was growing. The DOT were clearly unnerved – when over 3,000 people signed a pledge to use direct action to prevent a road being driven through London's last ancient woodland, Oxleas Wood, the DOT hastily withdrew the scheme. In the second half of 1993, the protests spread across the UK – from Skye and Newcastle in the north to Norwich in East Anglia and the M11 in London. The press and the public were becoming increasingly disturbed by the DOT's planned £23 billion roads programme and the forecast doubling in traffic. Even the construction industry was questioning the wisdom of the DOT's plans; *Construction News* wrote 'Many contractors share the anti-road lobby's puzzlement at the apparent discrimination in favour of roads and against rail and public transport.'

The campaign broadens

By September 1993, Transport Secretary John McGregor was calling for 'greater realism about the government's road programme' and in March 1994 the DOT dropped a third of its scheduled road schemes. However, the protests have gained too much impetus to die out after a few concessions are granted and continue to spread. In Yeovil demonstrators successfully prevented the construction of a supermarket on a greenfield site, by squatting for 6 weeks on the roof of an empty supermarket inside the town, pointing out the ludicrousness of fresh construction. A planned supermarket outside Bristol attracted similar protests and there have been similar demonstrations in Glasgow, Lincoln, and the East Midlands.

What started out as a series of single-issue protests against road expansion schemes has now mushroomed into a more general revitalization of the radical green movement. The protesters' aim is not just to stop individual road or supermarket schemes, but to challenge the compulsion for unending economic growth and overdevelopment at the expense of nature. Their success in getting this message across and attracting a wide spectrum of support has been substantial and may lie in the graphic simplicity of their tactics. Faced with the image of an unarmed protester defying the might of a bulldozer, the public's sympathy rarely lies with the government.

Simon Fairlie *The Ecologist*

the dosage required to kill 50% of the test group is found. Animal tests have become a target for criticism by ⇨animal rights groups, and alternatives have been sought, including tests on human cells cultured in a test tube and on bacteria.

In Europe in 1990 4,365 animals were used for testing cosmetic products, out of a total of 276,674 animals used to test products other than pharmaceuticals. Including pharmaceuticals, a total of around 3.2 million animals were used in experimentation. The US Office of Technology Assessment estimates that around 1.6 million animals are used annually in government research laboratories, of which 90% are rats or mice.

toxic waste dumped ⇨hazardous substance.

toxin any chemical molecule that can damage the living body. In vertebrates, toxins are broken down by enzyme action, mainly in the liver.

traffic vehicles using public roads. In 1970 there were 100 million cars and lorries in use worldwide; in 1990 there were 550 million. One-fifth of the space in European and North American cities is taken up by cars. In 1989 UK road-traffic forecasts predicted that traffic demand would rise between 83% and 142% by the year 2025.

In 1988 there were 4,531 deaths from motor-vehicle traffic accidents in England and Wales. The number of cars on UK roads has risen from 2.5 million (1951) to 20 million (1991). The congestion they cause costs £2–15 billion per year. The government spends £6 billion per year on roads, and receives around £14 million from fuel tax and vehicle excise duty. Road transport is responsible for 20% of UK CO_2 emissions.

transhumance seasonal movement by pastoral farmers of their livestock between areas of different climate. There are three main forms: in *Alpine* regions, such as Switzerland, cattle are moved to high-level pastures in summer and returned to milder valley pastures in winter; in *Mediterranean* lands, summer heat and drought make it necessary to move cattle to cooler mountain slopes; in *W Africa*, the nomadic herders of the Fulani peoples move cattle south in search of grass and water in the dry season and north in the wet season to avoid the tsetse fly.

transuranic element or *transuranium element* chemical element with an atomic number of 93 or more – that is, with a greater number of protons in the nucleus than has uranium. All transuranic elements are radioactive. Neptunium and plutonium are found in nature; the others are synthesized in nuclear reactions.

tributyl tin (TBT) chemical used in antifouling paints on ships' hulls and other submarine structures to deter the growth of barnacles. The tin dissolves in sea water and enters the food chain; the use of TBT has therefore been banned in many countries, including the UK.

triple nose-leaf bat one of many threatened bats in Africa, *Triaenops persicus* is found scattered along much of the coastal regions of East Africa and faces threats from disturbance of the caves in which it breeds. Tourism development, resulting in disturbance to coral caves which the bats inhabit, is a particular problem.

trophic level the position occupied by a species (or group of species) in a ⇨food chain. The main levels are ***primary producers*** (photosynthetic plants), ***primary consumers*** (herbivores), and ***secondary consumers*** (carnivores).

tuatara very rare lizard-like reptile *Sphenodon punctatus*, found only on a few islands off New Zealand. It grows up to 70 cm/2.3 ft long, is greenish black, and has a spiny crest down its back. On the top of its head is the pineal body, or so-called 'third eye', linked to the brain, which probably acts as a kind of light meter. It is the sole survivor of the ancient reptilian order Rhynchocephalia.

tuna any of various large marine bony fishes of the mackerel family, especially the genus *Thunnus*, popular as food and game. Albacore *T. alalunga*, bluefin tuna *T. thynnus*, and yellowfin tuna *T. albacares* are commercially important.

Tuna fish gather in shoals and migrate inshore to breed, where they are caught in large numbers. The increasing use by Pacific tuna fishers of enormous driftnets, which kill dolphins, turtles, and other marine creatures as well as catching the fish, has caused protests by environmentalists; tins labelled 'dolphin-friendly' contain tuna not caught by driftnets.

tundra region of high latitude almost devoid of trees, resulting from the presence of permafrost. The vegetation consists mostly of grasses, sedges, heather, mosses, and lichens. Tundra stretches in a continuous belt across N North America and Eurasia.

The term was originally applied to the topography of part of N Russia, but is now used for all such regions.

turbine engine in which steam, water, gas, air, or nuclear energy is made to spin a rotating shaft by pushing on angled blades, like a fan. Turbines are among the most powerful machines. Steam turbines are used to drive generators in power stations and ships' propellers; water turbines spin the generators in hydroelectric power plants; and gas turbines (as jet engines) power most aircraft and drive machines in industry. See also ⇨wind turbine.

turtle freshwater or marine reptile whose body is protected by a shell. Turtles are related to tortoises, and some species can grow to a length of up to 2.5 m/8 ft. Turtles often travel long distances to lay their eggs on the beaches where they were born, and many species have suffered through destruction of their breeding sites, often for tourist developments, as well as being hunted for food and their shell.

The hawksbill turtle, *Eretmochelys imbricata*, is now endangered, mainly through being hunted for its shell, which provides 'tortoise-shell'.

TVP (abbreviation for *texturized vegetable protein*) meat substitute usually made from soya beans. In manufacture, the soya-bean solids (what remains after oil has been removed) are ground finely and mixed with a binder to form a sticky mixture. This is forced through a spinneret and extruded into fibres, which are treated with salts and flavourings, wound into hanks, and then chopped up to resemble meat chunks.

U

uakari any of several rare South American monkeys of the genus *Cacajao*. There are three species. They have bald faces and long fur. About 55 cm/1.8 ft long in head and body, and with a comparatively short 15 cm/6 in tail, they are good climbers, remaining in the tops of the trees in swampy forests and feeding largely on fruit. The black uakari is in danger of extinction because it is found in such small numbers already, and the forests where it lives are fast being destroyed.

Ukraine *still struggling with the clean-up operation following the ⇨Chernobyl disaster 1986 which contaminated over 10% of the country, but despite this still depends on similar reactors for much of its power supply. As with many former Soviet republics, heavy industrialization under inadequate controls mean that industrial and chemical pollution is severe. The Dniester River, which supplies Odessa's drinking water, is heavily contaminated by industrial effluent and Kiev's water supply has been found to contain ⇨dioxins.*

UNEP acronym for *United Nations Environmental Programme*.

United Kingdom *an estimated 67% (the highest percentage in Europe) of forests have been damaged by acid rain and the UK is responsible for 9–12% of sulphur deposition in Norway. The UK has problems with water pollution, particularly in coastal areas, and many of the country's beaches have been condemned as unfit by the EC. It is one of the major polluters of the North Sea, partly due to accidents such as the Braer ⇨oil spill 1993, but also as a result of industrial and domestic pollution. In 1993, stratospheric ozone levels over Scotland were recorded at a record low of 23% below normal.*

United States of America *the USA is the largest contributor to the worldwide emission of* ⇨*greenhouse gases of any nation, releasing more than 20% of carbon dioxide emissions annually. Air pollution has become an increasing concern over recent years as many cities have been effected by* ⇨*smog as a result of vehicle emissions. This has led to measures such as the* ⇨*Clean Air Act and similar market-driven environmental measures to reduce pollution and increase energy conservation. The USA produces the world's largest quantity of municipal waste per person (850 kg/1,900 lb).*

unleaded petrol petrol manufactured without the addition of anti-knock. It has a slightly lower octane rating than leaded petrol, but has the advantage of not polluting the atmosphere with lead compounds. Many cars can be converted to running on unleaded petrol by altering the timing of the engine, and most new cars are designed to do so. Cars fitted with a ⇨catalytic converter must use unleaded fuel.

The use of unleaded petrol has been standard in the USA for some years, and is increasing in the UK (encouraged by a lower rate of tax than that levied on leaded petrol). In 1987 only 5% of petrol sold in the UK was unleaded; by 1992 this had risen to 45%. Between 1988 and 1990 UK lead emission fell by 30%.

uranium hard, lustrous, silver-white, malleable and ductile, radioactive, metallic element of the actinide series, symbol U, atomic number 92, relative atomic mass 238.029. It is the most abundant radioactive element in the Earth's crust, its decay giving rise to essentially all radioactive elements in nature; its final decay product is the stable element lead. Uranium combines readily with most elements to form compounds that are extremely poisonous. The chief ore is pitchblende.

Small amounts of certain compounds containing uranium have been used in the ceramics industry to make orange-yellow glazes and as mordants in dyeing; however, this practice was discontinued when the dangerous effects of radiation became known.

Uranium is one of three fissile elements (the others are thorium and plutonium). It was long considered to be the element with the highest atomic number to occur in nature. The isotopes U-238 and U-235 have been used to help determine the age of the Earth. Uranium-238, which comprises about 99% of all naturally occurring uranium, has a half-life of 4.51×10^9 years. Because of its abundance, it is the isotope from which fissile plutonium is produced in breeder ⇨nuclear reactors. The fissile isotope U-235 has a half-life of 7.13×10^8 years and comprises about 0.7% of naturally occurring uranium; it is used directly as a fuel for nuclear reactors and in the manufacture of nuclear weapons.

Many countries mine uranium; large deposits are found in Canada, the USA, Australia, and South Africa.

urban ecology study of the ecosystems, animal and plant communities, soils and microclimates found within an urban landscape. Although towns and cities seem at first sight to be entirely unnatural, a surprising amount of wildlife is able to exist among buildings and roads. While parks are important for many organisms, especially the song birds, birds of prey (the kestrel being a notable example) apparently find ample food in the wasteland around estates and offices. Mammals, including foxes and even badgers, are regular visitors, especially if there is an undisturbed corridor penetrating into the town, such as a disused railway.

urban sprawl outward spread of built-up areas caused by their expansion. This is the result of urbanization. Unchecked urban sprawl may join cities into conurbations; ⇨green-belt policies are designed to prevent this.

Increased mobility and improved transport systems in the developed countries such as the UK have encouraged urban growth that is often linear in form.

V

vicuna ruminant mammal *Lama vicugna* of the camel family that lives in herds on the Andean plateau. It can run at speeds of 50 kph/30 mph. It has good eyesight, fair hearing, and a poor sense of smell. It was hunted close to extinction for its meat and soft brown fur, which was used in textile manufacture, but the vicuna is now a protected species and populations are increasing after strict conservation measures were introduced.

Vietnam the Vietnam War caused serious problems; land was contaminated by dioxins as a result of the use of weapons such as ⇨Agent Orange as a defoliant and an estimated 2.2 million hectares of forest were destroyed. The country's National Conservation Strategy is trying to replant 500 million trees each year. Overfishing and pollution pose a serious threat to the once plentiful marine life in Vietnam's coastal waters.

vivisection literally, cutting into a living animal the term is now generally used to refer to any experiment on a living animal. In the UK over 3 million animals, excluding those used in military experiments, are killed in laboratory experiments every year.

In the late 19th century, vivisection was transformed from an eccentric hobby into an academic discipline. However it was only after World War II that the commercial potential of vivisection was realized by companies producing a wide range of synthetic chemicals, including pharmaceuticals, household and agricultural chemicals, cosmetics, food additives, and industrial chemicals.

⇨Animal rights campaigners object to vivisection not only from the ethical standpoint that alleged benefits to humans do not justify the suffering of animals, but also on the grounds that vivisection is scientifically invalid and misleading as it means applying results obtained from animals to human conditions. Such objectors point to the many drugs that have proved harmful to humans but have been passed as safe in animal tests, such as Thalidomide, and drugs such as penicillin which has provided some of the greatest medical advances of the 20th century but is harmful to guinea pigs. There were also long delays in implementing government warnings on asbestos and tobacco while the toxicity of these substances was tested on animals, even though the weight of epidemiological human evidence already showed that they were dangerous.

Vivisectors defend the practice by saying that human life is worth more than that of animals and hence some degree of animal suffering is acceptable in order to benefit people. Against the criticisms of the scientific validity of the experiments it is argued that while there are obvious differences between humans and animals, some results can be applied across the species barrier. Scientists claim that experiments which utilize such results have led to important breakthroughs in medical technology, for example in the treatment of diabetics.

W

Waldsterben (German 'forest death') tree decline related to air pollution, common throughout the industrialized world. It appears to be caused by a mixture of pollutants; the precise chemical mix varies between locations, but it includes acid rain, ozone, sulphur dioxide, and nitrogen oxides. *Waldsterben* was first noticed in the Black Forest of Germany during the late 1970s, and is spreading to many Third World countries, such as China.

Despite initial hopes that Britain's trees had not been damaged, research has now shown them to be among the most badly affected in Europe. Only 6% of the trees in Britain were undamaged by pollution in 1991, and about 57% had lost more than a quarter of their leaves. Acid rain is the main cause of this damage.

walrus Arctic marine carnivorous mammal *Odobenus rosmarus* of the same family (Otaridae) as the eared seals. It can reach 4 m/13 ft in length, and weigh up to 1,400 kg/3,000 lb. It has webbed flippers, a bristly moustache, and large tusks. It is gregarious except at breeding time and feeds mainly on molluscs. The walrus, particularly the Alaskan walrus, has been hunted close to extinction for its ivory tusks, hide, and blubber.

Washington Convention alternative name for ⇨*CITES*, the international agreement that regulates trade in endangered species.

waste materials that are no longer needed and are discarded. Examples are household waste, industrial waste (which often contains toxic chemicals), medical waste (which may contain organisms that cause disease), and ⇨nuclear waste (which is radioactive). By ⇨recycling, some materials in waste can be reclaimed for further use. In 1990 the industrialized nations generated 2 billion tonnes of waste. In the USA,

40 tonnes of solid waste are generated annually per person, roughly twice as much as in Europe or Japan.

There has been a tendency to increase the amount of waste generated per person in industrialized countries, particularly through the growth in packaging and disposable products, creating a 'throwaway society'.

In Britain, the average person throws away about ten times their own body weight in household refuse each year. Collectively the country generates about 50 million tonnes of waste per year. In principle, over 50% of UK household waste could be recycled, although less then 5% is currently recovered.

waste disposal depositing waste. Methods of waste disposal vary according to the materials in the waste and include incineration, burial at designated sites, and dumping at sea. Organic waste can be treated and reused as fertilizer (see ⇨sewage disposal). ⇨Nuclear waste and ⇨toxic waste are usually buried or dumped at sea, although this does not negate the danger.

Waste disposal is an increasing problem in the late 20th century. Environmental groups, such as Greenpeace and Friends of the Earth, are campaigning for more recycling, a change in lifestyle so that less waste (from packaging and containers to nuclear materials) is produced, and safer methods of disposal. Although incineration cuts down on ⇨landfill and can produce heat as a useful by-product it is still a wasteful method of disposal in comparison with recycling. For example, recycling a plastic bottle saves twice as much energy as is obtained by burning it.

The USA burns very little of its rubbish as compared with other industrialized countries. Most of its waste, 80%, goes into landfills. Many of the country's landfill sites will have to close in the 1990s because they do not meet standards to protect ground water.

The industrial waste dumped every year by the UK in the North Sea includes 550,000 tonnes of fly ash from coal-fired power stations. The British government agreed 1989 to stop North Sea dumping from 1993, but dumping in the heavily polluted Irish Sea will continue. Industrial pollution is suspected of causing ecological problems, including an epidemic that killed hundreds of seals 1989.

The Irish Sea receives 80 tonnes of uranium a year from phosphate rock processing, and 300 million gallons of sewage every day, 80% of it untreated or merely screened. In 1988, 80,000 tonnes of hazardous waste were imported into the UK for processing, including 6,000 tonnes of ⇨polychlorinated biphenyls.

water H_2O liquid without colour, taste, or odour. It is an oxide of hydrogen. Water begins to freeze at 0°C/32°F, and to boil at 100°C/212°F. When liquid, it is virtually incompressible; frozen, it expands by 1/11 of its volume. At 4°C/39.2°F, one cubic centimetre of water has a mass of one gram; this is its maximum density, forming the unit of specific gravity. It has the highest known specific heat, and acts as an efficient solvent, particularly when hot. Most of the world's water is in the sea; less than 0.01% is fresh water.

Water covers 70% of the Earth's surface and occurs as standing (oceans, lakes) and running (rivers, streams) water, rain, and vapour and supports all forms of Earth's life.

Water makes up 60–70% of the human body or about 40 litres of which 25 are inside the cells, 15 outside (12 in tissue fluid, and 3 in blood plasma). A loss of 4 litres may cause hallucinations; a loss of 8–10 litres may cause death. About 1.5 litres a day are lost through breathing, perspiration, and faeces, and the additional amount lost in urine is the amount needed to keep the balance between input and output. People cannot survive more than five or six days without water or two or three days in a hot environment.

A family of two adults and two children uses approximately 200 litres per day (UK figures). The British water industry was privatized 1989, and in 1991 the UK was taken to court for failing to meet EC drinking-water standards on nitrate and pesticide levels.

waterborne disease disease associated with poor water supply. In the Third World four-fifths of all illness is caused by water-borne diseases, with diarrhoea being the leading cause of childhood death. Malaria, carried by mosquitoes dependent on stagnant water for breeding, affects 400 million people every year and kills 5 million. Polluted water is also a problem in industrialized nations, where industrial dumping of chemical, hazardous, and radioactive wastes causes a range of diseases from persistent headache to cancer.

water cycle or *hydrological cycle* in ecology, the natural circulation

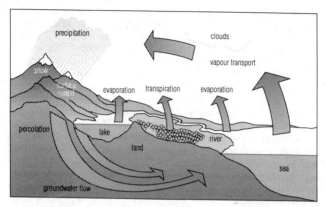

water cycle *About one-third of the solar energy reaching the Earth is used in evaporating water. About 380,000 cubic km/95,000 cubic mi is evaporated each year. The entire contents of the oceans would take about one million years to pass through the water cycle.*

of water through the ⇨biosphere. Water is lost from the Earth's surface to the atmosphere either by evaporation from the surface of lakes, rivers, and oceans, or through the transpiration of plants. This atmospheric water forms clouds that condense to deposit moisture on the land and sea as rain or snow. The water that collects on land flows to the ocean in streams and rivers.

water pollution any addition to fresh or sea water that disrupts biological processes or causes a health hazard. Common pollutants include nitrate, pesticides, and sewage (see ⇨sewage disposal), though a huge range of industrial contaminants, such as chemical by-products and residues created in the manufacture of various goods, also enter water – legally, accidentally, and through illegal dumping.

In the UK, water pollution is controlled by the National Rivers Authority (NRA) and, for large industrial plants, Her Majesty's Inspectorate of Pollution.

water supply distribution of water for domestic, municipal, or industrial consumption. Water supply in sparsely populated regions usually comes from underground water rising to the surface in natural springs, supplemented by pumps and wells. Urban sources are deep artesian wells, rivers, and reservoirs, usually formed from enlarged lakes or dammed and flooded valleys, from which water is conveyed by pipes, conduits, and aqueducts to filter beds. As water seeps through layers of shingle, gravel, and sand, harmful organisms are removed and the water is then distributed by pumping or gravitation through mains and pipes. Often other substances are added to the water, such as chlorine and fluorine; aluminium sulphate, a clarifying agent, is the most widely used chemical in water treatment. In towns, domestic and municipal (road washing, sewage) needs account for about 135 litres/30 gal per head each day. In coastal desert areas, such as the Arabian peninsula, desalination plants remove salt from sea water. The Earth's waters, both fresh and saline, have been polluted by industrial and domestic chemicals, some of which are toxic and others radioactive.

In 1980 the UN launched the 'Drinking Water Decade', aiming for cleaner water for all by 1990. However, in 1994 it is estimated that approximately half of all people in the developing world do not have safe drinking water.

In 1989 the regional water authorities of England and Wales were privatized to form 10 water and sewerage companies. Following public concern that some of the companies were failing to meet EC drinking-water standards on nitrate and pesticide levels, the companies were served with enforcement notices by the government Drinking Water Inspectorate.

wave power power obtained by harnessing the energy of water waves. Various schemes have been advanced since 1973, when oil prices rose dramatically and an energy shortage threatened. In 1974 the British engineer Stephen Salter developed the duck – a floating boom whose segments nod up and down with the waves. The nodding motion can be used to drive pumps and spin generators. Another device, developed in Japan, uses an oscillating water column to harness wave power. However, a major breakthrough will be required if wave power is ever to contribute significantly to the world's energy needs.

weathering process by which exposed rocks are broken down on the spot by the action of rain, frost, wind, and other elements of the weather. It differs from ⇨erosion in that no movement or transportation of the broken-down material takes place.

Three types of weathering are recognized: physical weathering, ⇨chemical weathering, and ⇨biological weathering. They usually occur together.

weedkiller or *herbicide* chemical that kills some or all plants. Selective herbicides are effective with cereal crops because they kill all broad-leaved plants without affecting grasslike leaves. Those that kill all plants include sodium chlorate and ⇨paraquat; see also ⇨Agent Orange. The widespread use of weedkillers in agriculture has led to an increase in crop yield but also to pollution of soil and water supplies and killing of birds and small animals, as well as creating a health hazard for humans.

wetland permanently wet land area or habitat. Wetlands include areas of marsh, fen, ⇨bog, flood plain, and shallow coastal areas. Wetlands are extremely fertile. They provide warm, sheltered waters for fisheries, lush vegetation for grazing livestock, and an abundance of wildlife. Estuaries and seaweed beds are more than 16 times as productive as the open ocean.

The term is often more specifically applied to a naturally flooding area that is managed for agriculture or wildlife. A water meadow, where a river is expected to flood grazing land at least once a year thereby replenishing the soil, is a traditional example. In the UK, the Royal Society for the Protection of Birds (RSPB) manages 2,800 hectares of wetland, using sluice gates and flood-control devices to produce sanctuaries for wading birds and wild flowers.

Many wetlands are threatened by development – for example, the Camargue in France, which is being developed for rice cultivation.

whale any marine mammal of the order Cetacea, with front limbs modified into flippers and with internal vestiges of hind limbs. The order is divided into the *toothed whales* (Odontoceti) and the *baleen whales* (Mysticeti). The toothed whales include ⇨dolphins and porpoises, along with large forms such as sperm whales. The baleen whales, with plates of modified mucous membrane called baleen in the mouth, are all large in size and include finback and right whales. There were hundreds of thousands of whales at the beginning of the 20th century, but they have been driven close to extinction by hunting and pollution (see ⇨whaling).

The ***blue whale*** *Sibaldus musculus*, one of the finback whales (or rorquals), is 31 m/100 ft long, and weighs over 100 tonnes. It is the largest animal ever to inhabit the planet. It feeds on plankton, strained through its whalebone 'plates'. The common ***rorqual*** *Balaenoptera physalas* is slate-coloured, and not quite so large. Largest of the toothed whales, which feed on fish and larger animals, is the ***sperm whale*** *Physeter catodon*. The ***killer whale*** (orca) is a large member of the dolphin family (Delphinidae), and is often exhibited in oceanaria. Killer whales in the wild have 8–15 special calls, and each family group, or 'pod', has its own particular dialect: they are the first mammals known to have dialects in the same way as human language. Right whales of the family Ballaenidae have a thick body and an enormous head. They are regarded by whalers as the 'right' whale to exploit since they swim slowly and are relatively easy to catch. The ***Northern right whale*** is close to extinction – numbers have fallen to an estimated 350. The ***bowhead*** whale *Balaena mysticetus* has strongly curving upper jawbones supporting the plates of baleen with which it sifts planktonic crustaceans from the water. Averaging 15 m/50 ft long and 90 tonnes/100 tons in weight, these slow-moving, placid whales were once extremely common, but by the 17th century were already becoming scarce through hunting. Only an estimated 3,000 remain, and continued hunting by the Inuit may result in extinction.

Whales are extremely intelligent and have a complex communication system, known as 'songs' which has become seriously affected by noise pollution – the sound of the 'songs' can be muffled or distorted by the noise from commercial shipping. Mass strandings where whales swim onto a beach occur occasionally for unknown reasons; it may be connected to chemical or noise pollution. Group loyalty is strong, and whales may follow a confused leader to disaster. Marine pollution in the North Sea from the start of the twentieth century is resulting in whales being born with spinal deformities; it is not known what the full extent of the damage caused in the late twentieth century will be.

whaling: chronology

11th/12th century	Basque villagers began hunting right whales in the Bay of Biscay, severely depleting local stocks by the 13th century.
1610	Greenland bowhead whale fishery began. Stocks were reduced to the brink of extinction by the 19th century.
1789	The first commercial whalers entered the Pacific.
1848	Arctic bowhead whales discovered and quickly exploited; stocks were almost totally depleted by 1914.
1890	Californian grey whale believed to be extinct after 40 years of hunting; however, some individuals survived and are now protected.
1904	Whaling operations began in the Southern Ocean.
1925	First factory ships used by whalers, doing away with the need to return to shore-based stations and so increasing catches.
1946	International Whaling Commission (IWC) established.
1950s	First commercial whale-watching tours began, observing migrating grey whales off S California.
1960–61	All-time peak in whaling; 64,000 cetaceans caught worldwide.
1985	IWC officially banned all commercial whaling; however, whaling continued under the guise of 'scientific' expeditions.
1988–89	Boycott of Icelandic fishing products, in protest against the country's continuation of commercial whaling, cost the Icelandic fishing industry $30 million.
1993	Greenpeace conducted a boycott campaign against Norwegian products to discourage Norway's continued whaling.
1994	Japan and Norway continued their commercial whaling, claiming it was for scientific research. Russia revealed that the former Soviet Union fleet had secretly continued large-scale commercial whaling for 40 years, without declaring catch numbers to IWC. Estimates of whale numbers worldwide have to be entirely revised.

whaling the hunting of whales, largely discontinued 1986. Whales are killed for whale oil (made from the thick layer of fat under the skin called 'blubber'), used for food and cosmetics; for the large reserve of oil in the head of the sperm whale, used in the leather industry; and for *ambergris*, a waxlike substance from the intestines, used in making perfumes. There are synthetic substitutes for all these products. Whales are also killed for their meat, which is eaten by the Japanese and was also used as pet food in the USA and Europe.

The International Whaling Commission (IWC), established 1946, failed to enforce quotas on whale killing until world concern about the possible extinction of the whale mounted in the 1970s. By the end of the 1980s, 90% of blue, fin, humpback, and sperm whales had been wiped out. Low reproduction rates mean that protected species are slow to recover. After 1986 only Iceland, Japan, Norway, and the USSR continued with limited whaling for 'scientific purposes', but Japan has been repeatedly implicated in commercial whaling, and pirates also operate. In 1990 the IWC rejected every proposal from Japan, Norway, and the USSR for further scientific whaling. Norway, Greenland, the Faroes, and Iceland formed a breakaway whaling club. In 1991 Japan held a 'final' whale feast before conforming to the regulations of the International Whaling Commission. In 1992 the IWC established the Revised Management Procedure (RMP), a new basis for regulating the exploitation of whales.

white-collared mangabey monkey of the genus *Cerocebus*. It lives in troops of about 20 in forest areas of midwest Africa. *C torquatus* is threatened mainly by forest destruction but is also hunted for food in some places and in Sierra Leone it is killed as a pest because troops sometimes raid crops.

wilderness area of uninhabited land which has never been disturbed by humans, which is usually located some distance from towns and cities. In the USA wilderness areas are specially designated by Congress and protected by federal agencies.

wildlife trade international trade in live plants and animals, and in wildlife products such as skins, horns, shells, and feathers. The trade has made some species virtually extinct, and whole ecosystems (for example, coral reefs) are threatened. Wildlife trade is to some extent regulated by ⇨CITES (Convention on International Trade in Endangered Species).

Species almost eradicated by trade in their products include many of the largest whales, crocodiles, marine turtles, and some wild cats. Until recently, some 2 million snake skins were exported from India every year. Populations of black rhino and African elephant have collapsed because of hunting for their horns and tusks (⇨ivory), and poaching

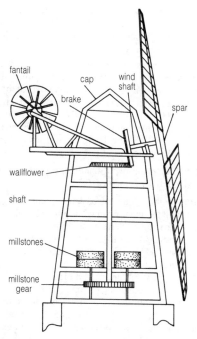

A simple tower windmill. The top cap revolves, blown by the wind catching the fantail, so that the large wooden sail always points into the wind. Power is transmitted to the grindstones by a series of shafts and gears.

remains a problem in cases where trade is prohibited.

wind farm large area of land covered by windmills or wind turbines, used for generating electrical power. A wind farm at Altamont Pass, California, USA, consists of 300 wind turbines. To produce 1,200 megawatts of electricity (an output comparable with that of a nuclear power station), a wind farm would need to occupy around 370 sq km/143 sq mi.

In the UK the first commercial windfarm, near Camelford, Cornwall, began to generate electricity in 1992.

wind power the harnessing of wind energy to produce power. The wind has long been used as a source of energy: sailing ships and windmills are ancient inventions. After the energy crisis of the 1970s ⇨wind turbines began to be used to produce electricity on a large scale.

By the year 2000, 10% of Denmark's energy is expected to come from wind power. Denmark now supplies 1.5% of its energy from wind energy alone. Since 1979 the Danish government has been giving grants of up to 30% of the cost of turbines.

Windscale former name of ⇨Sellafield, a nuclear power station in Cumbria, England. It was renamed 1971.

wind turbine windmill of advanced aerodynamic design connected to an electricity generator and used in wind-power installations. Wind turbines can be either large propeller-type rotors mounted on a tall tower, or flexible metal strips fixed to a vertical axle at top and bottom. In 1990, over 20,000 wind turbines were in use throughout the world, generating 1,600 megawatts of power.

The world's largest wind turbine is on Hawaii, in the Pacific Ocean. It has two blades 50 m/160 ft long on top of a tower 20 storeys high. An example of a propeller turbine is found at Tvind in Denmark and has an output of some 2 megawatts. Other machines use novel rotors, such as the 'egg-beater' design developed at Sandia Laboratories in New Mexico, USA.

Worldwide, wind turbines on land produce only the energy equivalent of a single nuclear power station, such as the Sizewell nuclear station. In Britain, 300 turbines, mainly in Cornwall, the northwest, and Wales, provide enough power for 60,000 homes.

The largest wind turbine on mainland Britain is at Richborough on the Kent coast. The three-bladed turbine, which is 35 m/115 ft across, produces 1 megawatt of power. Britain's largest vertical-axis wind turbine has two 24-m/80-ft blades and began operating in Dyfed, Wales, 1990.

wolf any of two species of large wild dogs of the genus *Canis*. The grey or timber wolf *C. lupus*, of North America and Eurasia, is highly social, measures up to 90 cm/3 ft at the shoulder, and weighs up to 45 kg/100 lb. It has been greatly reduced in numbers except for isolated wilderness regions.

The red wolf *C. niger*, generally more slender and smaller (average weight about 15 kg/35 lb) and tawnier in colour, may now be extinct in the wild. It used to be restricted to S central USA. Wolves were driven to extinction by hunting in England at the end of the 13th century, and in Scotland by the 17th century.

wolverine largest land member *Gulo gulo* of the weasel family (Mustelidae), found in Europe, Asia, and North America. It is stocky in build, about 1 m/3.3 ft long. Its long, thick fur is dark brown on the back and belly and lighter on the sides. It covers food that it cannot eat with an unpleasant secretion. Destruction of habitat and trapping for its fur have greatly reduced its numbers.

woodpecker bird of the family Picidae, which drills holes in trees to obtain insects. There are about 200 species worldwide. The largest of these, the imperial woodpecker *Campephilus imperialis* of Mexico, is very rare and may already be extinct.

World Wide Fund for Nature (WWF, formerly the *World Wildlife Fund*) international organization established 1961 to raise funds for conservation by public appeal. Projects include conservation of particular species, for example, the tiger and giant panda, and special areas, such as the Simen Mountains, Ethiopia.

The WWF has been criticized for investing in environmentally destructive companies, but the organization announced that this would cease. In 1990, it had 3.7 million members in 28 countries and an annual income of over £100 million. Its headquarters are in Gland, Switzerland.

World Wildlife Fund former and US name of the *World Wide Fund for Nature*.

WWF abbreviation for ⇨*World Wide Fund for Nature* (formerly World Wildlife Fund).

Y

yak species of cattle *Bos grunniens*, family Bovidae, which lives in wild herds at high altitudes in Tibet. It stands about 2 m/6 ft at the shoulder and has long shaggy hair on the underparts. It has large, upward-curving horns and humped shoulders. It is in danger of becoming extinct.

Yemen once known as Arabia Felix because of its fertility, large areas of the country now face land degradation and soil erosion due to careless agricultural activity, especially improper irrigation. Water supply is a major problem, with serious depletion of water tables by use of groundwater to supplement rainfall for drinking water and irrigation.

Z

Zaire *although Zaire has copious forest areas, including the largest tropical forest area in the world outside Brazil, poor management has meant that deforestation is still a problem. More serious is the threat to wildlife – poaching has taken a severe toll on several endangered species such as elephants and rhinoceroses. Unless urgent conservation measures are taken, it is possible that large mammals will soon only exist in national parks.*

---

zebra, Cape mountain zebra subspecies *equus zebra zebra*, confined to South Africa. It almost became extinct in the 1940s, and in 1993 had a population of only 450, despite attempts at conservation. The main population is in Mountain Zebra Park in the east of the country, although some zebras have been moved to other parks in an attempt to build up other viable breeding herds.

zoo abbreviation for *zoological gardens*, a place where animals are kept in captivity. Originally created purely for visitor entertainment and education, some Western zoos have attempted to transform their role into major centres for the breeding of endangered species of animals; a 1984 report identified 2,000 vertebrate species in need of such maintenance. The Arabian oryx has already been preserved in this way; it was captured 1962, bred in captivity, and released again in the desert 1972, where it has flourished. However, only 13 such 'captive breeding' schemes have actually been successful; many others have resulted only in further depletion of already low populations in the wild.

Zoos are criticized for keeping animals in unnatural conditions which causes stress and in some cases leads to psychosis. Apart from such allegations of cruelty, zoos are also criticized for putting additional pressure on endangered species for the sake of entertainment. For example, chimpanzees and gorillas displayed in zoos are often taken from the wild as infants which involves killing any other members of the social group present as they will attempt to defend their young. It is also claimed that zoos are of little more use educationally than films or pictures as the animals are seen in an entirely unnatural environment and therefore do not behave naturally.

London Zoo currently houses some 8,000 animals of over 900 species. Threatened by closure 1991 because of falling income, the zoo was given a one-year extension to July 1992; the decision to close it was reversed Sept 1992 and it is now planned to transform the zoo into a conservation park, with Whipsnade as the national collection of animals. In 1991 the number of animals in Britain's zoos totalled 35,000.

Appendices

air pollution

pollutant	sources	effects
sulphur dioxide SO_2	oil, coal combustion in power stations	acid rain formed, which damages plants, trees, buildings, and lakes; exacerbates asthma and causes irritation to eyes, nose, and throat
oxides of nitrogen NO, NO_2	high-temperature combustion in cars, and to some extent power stations	acid rain formed; exacerbates asthma, causes irritation of lung tissue, increases susceptibility to viral attack
lead compounds	from leaded petrol used by cars	slows development of neural tissue in children
carbon dioxide CO_2	oil, coal, petrol, diesel combustion	greenhouse effect
carbon monoxide CO	limited combustion of oil, coal, petrol, diesel fuels	leads to photochemical smog in some areas; deprives body of oxygen by combining with haemoglobin, causing headaches and drowsiness and can be fatal at high concentration
nuclear waste	nuclear power plants, nuclear weapon testing, war	radioactivity, contamination of locality, cancers, mutations, death

green movement: chronology

1798	Thomas Malthus's *Essay on the Principle of Population* published setting out the idea that humans are also bound by ecological constraints.
1824	Society for the Prevention of Cruelty to Animals founded.
1864	George Marsh's *Man and Nature* was the first comprehensive study of humans' impact on the environment.
1865	Commons Preservation Society founded, raising the issue of public access to the countryside, taken further by the mass trespasses of the 1930s.
1872	Yellowstone national park created in the USA; a full system of national parks was established 40 years later.
1893	National Trust founded in the UK to buy land in order to preserve places of natural beauty and cultural landmarks.
1930	Chlorofluorocarbons invented; they were hailed as a boon for humanity as they were not only cheap and nonflammable but were also thought not to be harmful to the environment.
1934	Drought in the USA exacerbated soil erosion, causing the 'Dust Bowl Storm', during which some 350 million tons of topsoil were blown away.
1948	United Nations created special environmental agency, the International Union for the Conservation of Nature.
1952	Air pollution caused massive smog in London, killing some 4,000 people and leading to Clean Air legislation.
1968	Garret Hardin's essay *The Tragedy of the Commons* challenged individuals to recognize their personal reponsibility for environmental degradation as a result of lifestyle choices.
1969	Friends of the Earth launched in USA as a more dynamic breakaway group from increasingly conservative Sierra Club; there was an upsurge of more radical active groups within the environmental movement over the following years.
1972	*Blueprint for Survival*, a detailed analysis of the human race's ecological predicament and proposed solutions, published by Teddy Goldsmith and others from the *Ecologist* magazine.
1974	First scientific warning of serious depletion of protective ozone layer in upper atmosphere by CFCs.
1980	US president Jimmy Carter commissioned Global 2000 report, reflecting entry of environmental concerns into mainstream political issues.
1983	German Greens (Die Grünen) won 5% of vote, giving them 27 seats in the Bundestag.
1988	NASA scientist James Hansen warned US Congress of serious danger of global warming: 'The greenhouse effect is here'.

green movement: chronology (continued)

1985	Greenpeace boat *Rainbow Warrior* sunk by French intelligence agents while in a New Zealand harbour during a protest against French nuclear testing in the S Pacific. One crew member was killed.
1989	European elections put green issues firmly on political agenda as Green parties across Europe attracted unprecedented support; especially in the UK, where the Green Party received some 15% of votes cast (though not of seats).
1989	*The Green Consumer Guide* published in the UK, one of many such books worldwide advocating 'green consumerism'.
1991	The Gulf War had massive environmental consequences, primarily as a result of the huge quantity of oil discharged into the Gulf from Kuwait's oilfields.
1992	United Nations Earth Summit in Rio de Janeiro aroused great media interest but achieved little progress in tackling difficult global environmental issues as many nations feared possible effects on trade.
1994	Anti-road protests in the UK reached new height with 'Battle of Wanstonia' as green activists occupied buildings and trees in Wanstead, E London, in attempt to halt construction of M11 motorway.

E numbers

a selection of food additives authorized by the European Commission

number	name	typical use
COLOURS		
E102	tartrazine	soft drinks
E104	quinoline yellow	
E110	sunset yellow	biscuits
E120	cochineal	alcoholic drinks
E122	carmoisine	jams and preserves
E123	amaranth	
E124	ponceau 4R	dessert mixes
E127	erythrosine	glacé cherries
E131	patent blue V	
E132	indigo carmine	
E142	green S	pastilles
E150	caramel	beers, soft drinks, sauces, gravy browning
E151	black PN	
E160 (b)	annatto; bixin; norbixin	crisps
E180	pigment rubine (lithol rubine BK)	
ANTIOXIDANTS		
E310	propyl gallate	vegetable oils, chewing gum
E311	octyl gallate	
E312	dodecyl gallate	
E320	butylated hydroxynisole (BHA)	beef stock cubes; cheese spread
E321	butylated hydroxytoluene (BHT)	chewing gum
EMULSIFIERS AND STABILIZERS		
E407	carageenan	quick-setting jelly mixes, milk shakes
E413	tragacanth	salad dressings; processed cheese
PRESERVATIVES		
E210	benzoic acid	
E211	sodium benzoate	beer, jam, salad cream, soft drinks, fruit pulp
E212	potassium benzoate	fruit-based pie fillings, marinated herring and mackerel

E numbers (continued)

number	name	typical use
E213	calcium benzoate	
E214	ethyl para-hydroxy-benzoate	
E215	sodium ethyl para-hydroxy-benzoate	
E216	propyl para-hydroxy-benzoate	
E217	sodium propyl para-hydroxy-benzoate	
E218	methyl para-hydroxy-benzoate	
E220	sulphur dioxide	
E221	sodium sulphate	dried fruit, dehydrated vegetables, fruit juices and syrups, sausages, fruit-based dairy desserts, cider, beer, and wine; also used to prevent browning of peeled potatoes and to condition biscuit doughs
E222	sodium bisulphite	
E223	sodium metabisulphite	
E224	potassium metabisulphite	
E226	calcium sulphite	
E227	calcium bisulphite	
E249	potassium nitrite	
E250	sodium nitrite	bacon, ham, cured meats, corned beef and some cheeses
E251	sodium nitrate	
E252	potassium nitrate	
OTHERS		
E450 (a)	disodium dihydrogen diphosphate	butters, sequestrants, emulsifying salts, stabilizers, texturizers
	trisodium diphosphate	
	tetrasodium diphosphate	
E450 (b)	tetrapotassium diphosphate pentasodium triphosphate pentapotassium triphosphate	raising agents, used in whipping cream, fish and meat products, bread, processed cheese, canned vegetables

deforestation during the 1980s

continent	total land area (km²)	forest area 1980 (km²)	forest area 1990 (km²)	annual deforestation 1981–90 (km²)	deforestation rate 1981–90 (%)
Latin America	16,756,000	9,229,000	8,399,000	84,000	0.9
Asia	8,966,000	3,108,000	2,748,000	35,000	1.2
Africa	22,433,000	6,504,000	6,001,000	51,000	0.8
Total	48,155,000	18,841,000	17,148,000	170,000	0.9

Source: FAO 1991 Second Interim Report on the State of Tropical Forests by Forest Resources Assessment 1990 Project Tenth World Forestry Congress, September 1991, Paris, France.

endangered species

species	observation
plants	one-quarter of the world's plants are threatened with extinction by the year 2010
amphibians	worldwide decline in numbers; half of New Zealand's frog species are now extinct
birds	three-quarters of all bird species are declining or threatened with extinction
carnivores	almost all species of cats and bears are declining in numbers
fish	one-third of North American freshwater fish are rare or endangered; half the fish species in Lake Victoria, Africa's largest lake, are close to extinction due to predation by the introduced Nile perch
invertebrates	about 100 species are lost each day due to deforestation; half the freshwater snails in the southeastern USA are now extinct or threatened; one-quarter of W German invertebrates are threatened
mammals	half of Australia's mammals are threatened; 40% of mammals in France, the Netherlands, Germany, and Portugal are threatened
primates	two-thirds of primate species are threatened
reptiles	over 40% of reptile species are threatened

nuclear electricity supplied worldwide

year	% of total
1960	0.1
1965	0.7
1970	1.6
1975	5.5
1980	8.3
1985	15.3
1991	17.0

Source: International Atomic Energy Authority

major oil spills

year	place	source	quantity	
			tonnes	*litres*
1967	off Cornwall, England	*Torrey Canyon*	107,100	
1968	off South Africa	*World Glory*		51,194,000
1972	Gulf of Oman	*Sea Star*	103,500	
1977	North Sea	Ekofisk oilfield		31,040,000
1978	off France	*Amoco Cadiz*	200,000	
1979	Gulf of Mexico	Ixtoc 1 oil well	535,000	
1979	off Trinidad and Tobago	collision of *Atlantic Empress* and *Aegean Captain*	270,000	
1983	Persian Gulf	Nowruz oilfield	540,000	
1983	off South Africa	*Castillo de Beliver*	225,000	
1989	off Alaska	*Exxon Valdez*		40,504,000
1991	Persian Gulf	oil wells in Kuwait and Iraq		
1993	Shetland Islands, Scotland	*Braer*	85,000	

state support of railways 1988

country	total £ million	per head of population
UK	634.4	£11.43
France	3,005	£53.79
West Germany	3,509	£57.34
Italy	4,007	£69.75

international railways 1988

country	total length (km)	passenger km/year (millions)
USSR	247,000	
USA	225,000	19,151
France	34,600	59,732
West Germany	27,400	39,174
UK	16,600	33,140
Spain	12,700	15,394
Portugal	3,600	5,907
ranking 15th in the world		

British Rail passenger use

	journeys (millions)	passenger km/year (millions)
1965	865	30,116
1975	730	30,300
1985	697	29,700

recycling

Levels of recycling of packaging waste, selected European countries, 1980–1990 (percentage of total consumption)

	paper and cardboard		glass		aluminium	
	1980	1991	1980	1991	1989	1990
EC Members						
Belgium	14.7		42.0	55.0	59.0	
Denmark	25.6	29.7*	53.9		35.0	40.0
France	37.0	46.7*	26.0	41.0		41.0
Germany	33.9	43.0	35.5	63.0	3.0	54.0
Greece				22.0	21.0	16.0
Ireland	15.0		7.0	23.0	2.0	19.0
Italy			25.0	53.0	8.0	48.0
Luxembourg						
Netherlands	45.5	66.1*	53.0	70.0		66.0
Portugal	38.0	10.0	14.0	23.0		
Spain	38.1	44.1	64.1	22.0	27.0	27.0
United Kingdom	29.0	59.0	12.0	21.0		21.0
EFTA members						
Austria		39.0*	38.0	60.0	3.0	60.0
Finland	30.0	5.2*	21.0	31.0		46.0
Iceland						
Liechtenstein						
Norway	21.9	23.2		22.0		34.0
Sweden	34.0		12.3	44.0	82.0	35.0
Switzerland	38.0		36.0	71.0	31.0	61.0

* 1990 figures used where 1991 figure unavailable
Source: OECD

water scarcity

region/country	per capita renewable water supplies		change
	1992	*2010*	
	(cubic metres per person)		*(%)*
Africa			
Algeria	730	500	−32
Botswana	710	420	−41
Burundi	620	360	−42
Cape Verde	500	290	−42
Djibouti	750	430	−43
Egypt	30	20	−33
Kenya	560	330	−41
Libya	160	100	−38
Mauritania	190	110	−42
Rwanda	820	440	−46
Tunisia	450	330	−27
Middle East			
Bahrain	0	0	0
Israel	330	250	−24
Jordan	190	110	−42
Kuwait	0	0	0
Qatar	40	30	−25
Saudi Arabia	140	70	−50
Syria	550	300	−45
United Arab Emirates	120	60	−50
Yemen	240	130	−46
Barbados	170	170	0
Belgium	840	870	+4
Hungary	580	570	−2
Malta	80	80	0
Netherlands	660	600	−9
Singapore	210	190	−10

whale catches by species, 1980–1991

	year					
	1980	*1981*	*1982*	*1983*	*1984*	*1985*
bowhead	34	28	19	18	25	17
bryde	970	648	802	697	709	357
fin	471	410	356	278	28l	219
gray	181	136	l68	171	169	170
humpback	18	14	16	16	15	8
minke	10,910	11,320	10,946	9,796	7,541	7,168
sei	103	100	71	100	95	38
sperm	2,092	1,452	621	414	463	400
total	14,779	14,108	12,999	11,490	9,298	8,377

	year					
	1986	*1987*	*1988*	*1989*	*1990*	*1991*
bowhead	28	31	29	26	44	47
bryde	317	317				
fin	85	89	77	82	19	16
grey	171	158	l51	180	163	⁻170
humpback	2	2	2	2	1	2
minke	5,875	1,040	389	422	429	414
sei	40	20	10	2		
sperm	211	211	8			
total	6,729	1,868	586	714	656	649

Figures are from whale catch figures declared to the IWC by the following countries: Denmark (including Greenland), Iceland, Norway, Spain, Portugal. St Vincent & Grenadines, Japan, Korea.Taiwan, USA (including Alaskan Eskimos), USSR (including Siberian Aleuts), Brazil, Chile, Peru, Philippines, Indonesia, and Italy.

Trends in emissions of CO_2 from industrial sources in major world regions, 1950–1990

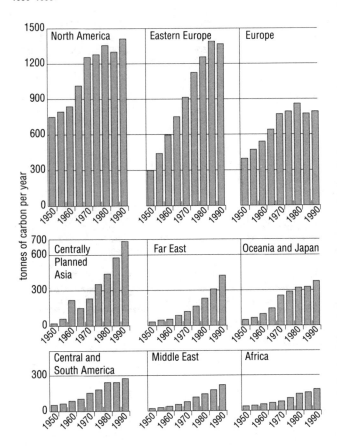

European emissions CO₂ 1880–1990

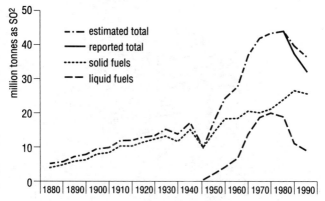

Ozone concentration isopleths over the southern polar regions

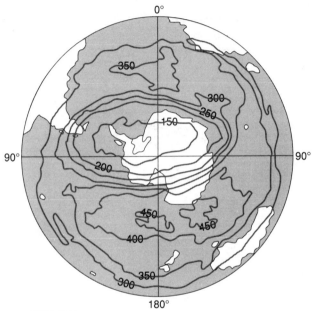

source: WMO, 1993a

Ozone levels in the Antarctic, 1957–1990

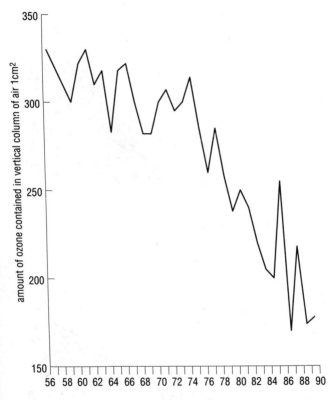

Source: UKSORG

World releases of CFC-11 and CFC-12 1931–89

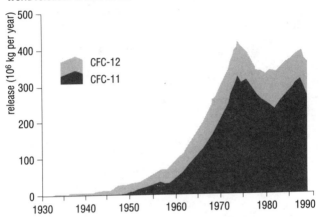

Estimated emissions fo halon (H-1211 and H-1301) worldwide, 1960–90

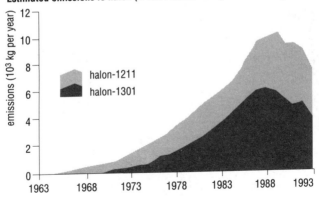

Threats to mammal species worldwide

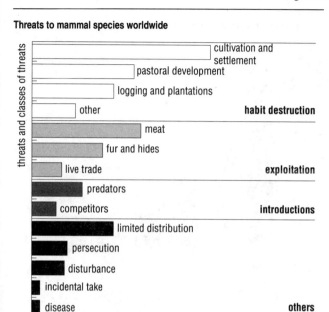

Per cent of known species classed as endangered

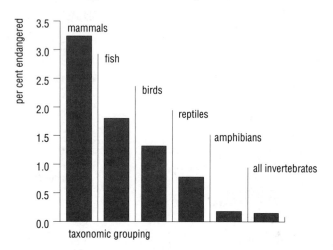

The proportion of whales caught during 1980–91 by the main whaling nations

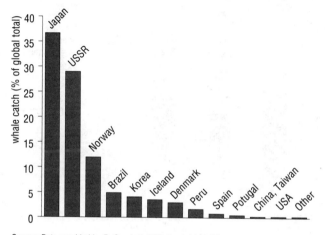

Source: Data provided by R. Gambell, IWC, Cambridge, UK

Pesticide poisoning of animals in the UK, 1988–1991

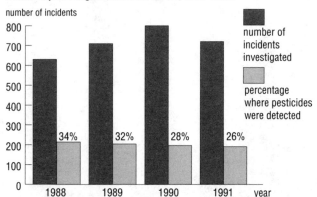

Percentage of incidents where pesticides were detected, 1991

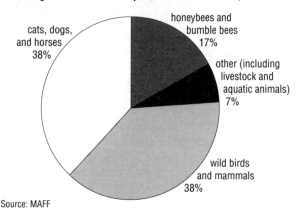

Source: MAFF

Waste arising by sector UK (total waste 400 million tonnes annually)

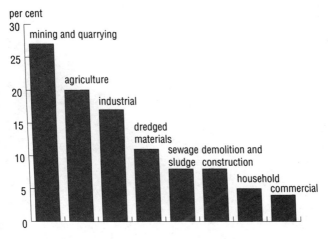

Source: MAFF